Introduction to Food Science and Technology

SECOND EDITION

FOOD SCIENCE AND TECHNOLOGY

A SERIES OF MONOGRAPHS

Series Editors

A complete list of the books in this series appears at the end of the volume.

Introduction to
Food Science and Technology
SECOND EDITION

George F. Stewart

Department of Food Science and Technology
University of California, Davis

Maynard A. Amerine

Department of Viticulture and Enology
University of California, Davis

1982

ACADEMIC PRESS
A Subsidiary of Harcourt Brace Jovanovich, Publishers

New York   London
Paris   San Diego   San Francisco   São Paulo   Sydney   Tokyo   Toronto

ACADEMIC PRESS, INC.
111 Fifth Avenue, New York, New York 10003

United Kingdom Edition published by
ACADEMIC PRESS, INC. (LONDON) LTD.
24/28 Oval Road, London NW1 7DX

Library of Congress Cataloging in Publication Data

Stewart, George Franklin, Date
    Introduction to food science and technology.

    (Food science and technology)
    Includes bibliographical references and index.
    1. Food industry and trade. I. Amerine, M. A.
(Maynard Andrew), Date   . II. Title. III. Series.
TP370.S63 Date   664   82-6720
ISBN 0-12-670256-X   AACR2

On December 12, 1981, Professor George Stewart and I finished reading the final text of the second edition. A few days later the completed manuscript was sent to the publisher. On March 18, after a brief illness, Professor Stewart died. George had devoted much of his time and energy over the past two years to the completion of this book. Its merits are largely due to his careful editing of the manuscript. Any errors are more mine than his.

My personal debts to him are enormous. He had an encyclopedic knowledge of food science and technology, accumulated during his years at Iowa State University and the University of California at Davis, where he was chairman of three departments. He was unfailingly courteous and kindly, a gentleman and a scholar. We shall not see his like again soon.

Maynard A. Amerine

Contents

Contents

Preface

This second edition of *Introduction to Food Science and Technology* addresses the same subject matter as did the first, and it intends to serve the same audience. Thus, its primary objective is to acquaint college freshmen and sophomores with food science and technology and possibly to motivate them to consider the field as a career. This text is not designed to appeal to those primarily interested in topical debates about the American food industry and its products but rather to inform the student of our present scientific knowledge about foods and of the application of scientific principles to commercial food processing and preservation. We have, however, included a chapter examining consumer concerns and current controversies about food choices and health.

As part of the work of preparing this book, the authors surveyed academicians in American and Canadian university food science departments who have used the earlier edition, as well as others familiar with it. The excellent suggestions and usable ideas we received have helped us to revise and update substantially the material presented in this edition. We have also added three chapters to allow for expanded coverage of certain important topics.

As before, we have not tried to present the subject matter at an advanced technical level. However, since the book primarily concerns scientific topics, we have assumed a certain basic level of scientific understanding on the part of the reader. We would expect the student using this text to have had at least college-preparatory courses in mathematics, chemistry, physics, and biology. The student could have taken similar courses in the freshman collegiate year rather than in high school. Such preparation should make the scientific discussions presented in the book relatively easy to comprehend.

Food science and technology is a relatively new field. Only recently has it begun to achieve a degree of technical maturity in its development, but it is rapidly being recognized, nationally and internationally, as a viable, important, and necessary field of study for those who wish to pursue technical careers in food processing and preservation.

As the name implies, the field is multidisciplinary and encompasses both basic and applied subjects. Thus, food science deals with the study of certain facets of the physical, chemical, biological, and behavioral sciences as they influence the processing and preservation of food. Food technology, on the other hand, deals with the engineering and other scientific and technical problems involved in transforming edible raw materials and other ingredients into safe, pure, nutritious, and appetizing food products. In other words, food science is concerned with the basic scientific facts about food, whereas food technology is concerned with the commercial processing of raw materials into foods that meet human needs and wants.

The scientific study of foods and the processing and preservation of food in factories are not new activities, of course. However, over the past 50 to 100 years there has been a revolution in this field as American society has changed from a largely rural population, self-sufficient in food, to a nation of urban dwellers, highly dependent on others for food and other basic needs. For example, in the United States fewer than 5% of the population lives on farms and produces food for sale; yet this small group of people supplies almost all the food needed by the other 95%. In addition, the processing and preservation of that food is now carried out primarily in large factories and seldom in the home or in the small-scale establishments of past centuries.

This revolution in agriculture and in the other industries that produce raw materials and supplies for food manufacturing is feasible because of our advancing knowledge of food science and because of the engineering developments that have made possible the large-scale processing and preserving of food. Accordingly, food science and technology has developed over the past four or five decades into a large, scientifically well-grounded, and technically sophisticated specialty.

Academic training for technical careers in food science and technology requires a broad, in-depth education both in certain sciences and in selected engineering specialties. It is precisely because of the complex nature of food and its processing and the requirement for a rigorous scientific/technical training that food science and technology offers an exceptional opportunity and a real challenge to the bright applications-oriented science student seeking a rewarding career.

The basic purpose of this text, then, is to offer the prospective food science major a brief but comprehensive outline of the field. The authors have endeavored to suggest the breadth of the field and to show the importance of applying sound scientific principles to solving the problems of transforming raw materials into safe, pure, nutritious, and appealing food products. We would hope that the student who chooses this field as an academic major will find a rewarding career and will develop a respect for and pride in the work.

As noted above, this book was written primarily for the beginning college student. However, the authors hope that others, including members of the general public, will find that it provides an accurate picture of the technical nature of the modern food industry.

Chapter 1

Evolution of Food Processing and Preservation

Is there any thing whereof it may be said, See this is new? It hath been already of old time, which was before us.

Ecclesiastes 1:10

Early Developments

Primitive humans gathered food as early as 1 million years ago. They fed themselves by harvesting wild fruits and vegetables and catching small animals, insects, and fish. From earliest human history to the present, food gathering and processing have become more and more diversified and complex.

Pre-Neolithic Period

Peking man (possibly 250,000 years ago) used fire for cooking and hunted a variety of animals, including other humans. Cannibalism has been practiced by humans since the earliest days (and it still is in cases of extreme starvation— witness the Donner party in California in the last century and the Andean air crash in 1977). Fire not only kept people warm but also lighted their homes, protected them from wild animals, provided a community center, and profoundly modified their foods. Roasted meats had a different texture and flavor and spoiled less readily than uncooked meats. Cooking increased the nutritive value of foods and in some cases made them more digestible or chewable. It also killed potentially harmful microorganisms, destroyed some toxic chem-

1

icals, retarded decay, and made foods taste better and plesantly warm. (For a survey of the diets of early humans—vertebrates, including man, invertebrates, fungi, and plants—see Brothwell and Brothwell, 1969; good bibliography and illustrations.)

Paleolithic Period. During the Paleolithic period humans not only roasted food but developed grinding, pounding, and drying as methods of food processing (Table 1). Wendorf *et al.* (1979) discovered ground barley in Egypt dated as early as 18,000 B.C. These authors considered this a food-producing economy—possibly the earliest.

One aspect of food processing that is often neglected is processing, to an edible state, plant or animal products that have undesirable storage, taste, or toxic properties. Drying, salting, fermentation, cooking, smoking, crushing, pressing, grating, and pounding are processes that were developed to extend the storage life of many raw materials. Such procedures as cooking, baking, crushing, peeling, fermenting, pressing, grating and pounding were often preludes to drying (Yen, 1975). Drying was used in the arid Middle Eastern countries and also in other areas. In Polynesia, Micronesia, and Melanesia, drying , with or without prior smoking, is still widely practiced (Yen, 1975). Drying made possible food storage, and this provided food when crops were unavailable. It also allowed for increased populations, especially in environments of low natural carrying capacities (for example, in Oceania). Drying in the sun is still practiced, and dehydration, with or without reduced pressure, is at present a major food process (see p. 221–224).

Mesolithic Period. In the Mesolithic period hunting was man's predominant occupation. So desperate was he for food that even very large animals were hunted (Tannahill, 1973). The disappearance of some animal species appears to have been due to overhunting. Hunting is a precarious method of providing food regularly. A large area was needed to feed a family—21 square miles for the early Germanic tribes. In some cases even more territory may have been required. In some areas, collection of small animals for food seems to have been important. The supply of animals varied, and so periods of starvation resulted. Migration to other areas was often necessary to ensure productive hunting. Because of their migratory nature, hunting of many species was highly seasonal. The all-meat diet also was monotonous and the meat was difficult to keep. Stone (often flint), horn, and bone flakes were used to skin and cut up meat (Stanford *et al.,* 1981). Paleolithic man in northern Europe during the last Ice Age hunted big game almost exclusively: reindeer, horses, mammoths, bison. Fishing and bird hunting appear to have been of limited importance. Saffiro (1975) emphasizes the small amounts of vegetables consumed by Eskimos and of others living in extremely cold regions (daily caloric requirement of over 6000 calories!) without harmful effects.

Even before pottery-making developed, a new food process—boiling—was employed. When available, hot springs were used; more commonly, hot rocks were dropped into water. Tannahill (1973) emphasizes the use of the animal stomach as a cooking vessel. Smoking also developed as a method of food preservation that provided a new flavor. There appears to have been more storage of food in the Mesolithic period. Fish were probably also dried. The bow and arrow facilitated hunting. The search for food led to group organization, use of tools, sharing the catch, verbal communication, and finally to the division of labor and the assumption of specific roles.

Some animals were domesticated very early—dogs, sheep, goats, cattle and reindeer. It has been suggested that some of these aniamls may have been domesticated through the keeping of cult or totem animals in temples or sacred precincts.

Neolithic Period and Revolution

Prior to the agricultural revolution of 7,000 to 10,000 years ago, humans had been carnivorous for hundreds of thousands of years. Before that, humans had been omnivorous, eating meat and many plants. Presumably our inherent taste for sweets comes from eating fruits during this early period.

Since the agricultural revolution humans have returned to a more herbivorous diet. Worldwide, 21% of the food consumed is rice and 20% wheat! Primitive hunting-gathering societies did better in eating a variety of foods, and some still do. For example, primitive Gold Coast groups used 114 species of fruit, 46 of leguminous seeds, and 47 of greens (de Castro, 1952).

The Neolithic period marks the change from a food-gathering and hunting society to one of food production. The transition was gradual, starting at various times in different geographical areas, and was more or less complete in most, but not all, parts of the world by about 1000 B.C. For example, the Indians of northern California were exclusively food gatherers in the pre-Spanish period (Heizer and Elasser, 1980). Hunting, of course, continued to be an important food source in certain areas (as it still is today in areas where the reindeer are semidomesticated). Wild and domesticated animals were used simultaneously in early cultures (Bökönyi, 1975).

The cause of the Neolithic revolution is not known (Heiser, 1981). The known change to a warmer world climate about this time may have induced humans to leave their cave dwellings and settle in the open. The warmer climate may also have stimulated their interest in the domestication of plants and animals. At any rate, in the early Neolithic period beef cattle, buffalo, yak, banteng, and pigs were domesticated. Domesticated animals provided an easily available food source (including milk and eggs); they also supplied fertilizer, while their skins and feathers were used as clothing. The human population thereafter increased rapidly.

TABLE 1

Development of Agriculture, Food Processing, and Food Preservation

Period	Dates[a]	Agriculture	Foods	Processing and preservation techniques	Examples of science and technology
Upper Paleolithic	Before 15,000 B.C.	None; fishing	Eggs, fruit, nuts, seeds, roots, insects, fish, honey, small and large animals[b]	Roasting, pounding, drying, grinding, freezing(?)	Bags, baskets; cloth, stone, and bone implements; fish hooks, "made" fire, painting, sculpture, language
Mesolithic	15,000 B.C.	None	Great variety; stored wild fruits and berries	Dried fish, boiling, food storage, steaming(?)	Bow and arrow; dog, goat, reindeer, and sheep probably domesticated; clay-covered baskets
Neolithic	9000 B.C. or earlier	Seasonal culture of cereals, hoe culture, plowing, permanent fields	Domesticated animals,[c] milk, butter, cheese, gruel, beer, vinegar	Alcoholic fermentation, acetification, salting, baking, breadmaking, sieving, primitive pressing, seasoning	Pottery wheel; spinning, weaving; wood, flint, and bone sickles; saddle quern; mortar; fishing with hooks and nets

4

Bronze (cities)	3500 B.C.	Irrigation[d], horse- and ox-drawn plows, much local and long-distance trade, vegetative propagation, fruit grown, pruning	Soybeans, figs, rice, olives, olive oil, vegetables, lentils, cabbage, cucumbers, onions, pomegranates, dates, grapes, wine	Filtration, lactic acid fermentation, more types of flavoring, flotation, leavened bread, kneading, pickling, sausage making, frying, sophisticated and complicated pressing, clarification	Architecture, smelting; wheeled carts, ships, writing, bronze tools, mathematics, rotary millstones, bronze weapons, astronomy, shadufs, medicine, chemistry
Iron	1500 B.C.	Land and sea trade common; heavier plows	Apples, pears, cherries, spices, beans, artichokes, lettuce, sauces	Refinement of flavoring and of cookery	Pulleys, glass, improved and cheaper tools and weapons, currency
Roman	600 B.C.–400 A.D.	Reaping machines, legume rotation, plows on wheels, food trade	Sugarcane in West; asparagus, beets, oranges	Food adulteration common	Water mills, donkey mills, wooden cooperage

[a] The dates indicate only the beginnings in the main centers of origin. The items referred to appeared much later in other areas, may not have developed at all, or may even have retrogressed. Tasmania was discovered in 1642 and the evidence up to Cook's visit in 1777 was that the people of Tasmania had retrogressed from the Neolithic to Paleolithic period. Stone implements continued to be used long after the Paleolithic period. It is also important to remember that a food-processing operation may have originated in one region long before another. Bronze Age implements continued to be used for a long time into the Iron Age.

[b] Big-game hunts occurred in areas of cliffs about 400,000 B.C. when fire and axes and spears were used. Pit-hunting and use of knives appeared about 75,000 B.C.

[c] The order of domestication is unknown; however, goats, yak, buffalo, pigs, and cattle were domesticated early in this period, but not horses or camels. Horses, camels, asses, elephants, and poultry were domesticated toward the end of this period.

[d] Irrigation existed prior to 3500 B.C. but its widespread use about this time is believed to account for the spectacular increase in the population of Mesopotamia.

Cereals. The domestication of cereals—in which most historians assign women a major role—led to significant changes in the human life-style. The earliest cultivated cereals were emmer and einkorn wheat, barley, and rice; millet, oats, and buckwheat followed. The diet changed from a predominantly carnivorous one to a more balanced vegetable and meat diet; from a wholly nomadic life to an interest in the particular area where food was being produced; from a life of periodic food shortages to one of food surpluses; and finally, to a settled village life. More important, cereals contained carbohydrates, fats, proteins, minerals, and vitamins. They could also be used for animal feed and bedding, litter baskets, and so on; and production could be expanded as demand increased.

Cereal grains produced a huge yield from a single seed; they could be stored for several years and could be easily prepared for eating by roasting, removal of the husk, grinding or pounding, and soaking. The saddle quern or the mortar and pestle used for grinding cereals were characteristic of the Neolithic period (see Fig. 1). A sort of grain-paste was widely used; later it was baked on a hot stone. The Mexican tortilla and the Indian chapati are modern examples of such a product. Later the ground cereal was fermented to make beer, or fermented and baked into bread. Early agriculture soon developed weeding, fertilization, hoeing, and, most important, irrigation of crops.

The domestication of other plants, particularly root crops and corn, greatly expanded the food supply. Of the 3000 species of plants that have been used for food, only about 150 have been extensively cultivated. The major cultivated plants used today are rice, wheat, corn, sugarcane, sugar beet, potato, sweet potato, cassava, common bean, soybean, coconut, and banana.

The probable place of origin of various economically important plants is shown in Fig. 2 (see also Table 2). However, there is still much argument as to where agriculture first developed, the priority of tuber-based or cereal-based agriculture, and why the Old World emphasized domestic animal production more than the New World did (Reed, 1977). Even the place of

Fig. 1. Making spiced bread, which might be subsequently further baked for eating or, alternatively, soaked in water and fermented to make beer. From the right: the grain is dehusked in mortars, sieved, and ground on a quern. The group of women on the left then form it into cones of dough, which are baked on the fire in the center. The woman on the extreme left is coloring a cone of dough with a red pigment. From a tomb at Thebes, Egypt, about 1900 B.C. (From Singer *et al.*, 1954–1958.)

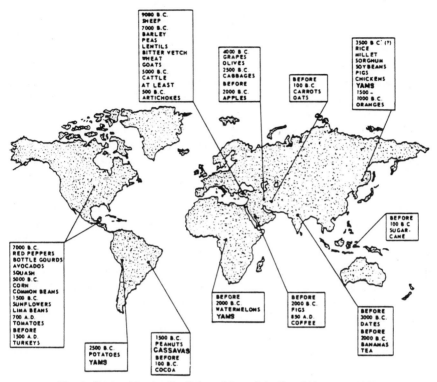

Fig. 2. Origin of foodstuffs. (Adapted from Sebrell and Haggerty, 1968.)

TABLE 2

**Probable Original Center of Distribution of
the Ancestors of Modern Economic Plants[a]**

Center of distribution	Species
Central Asia	Apple, barley, broad bean, carrot, celery, chick pea, cucumber, date, eggplant, grape, lentil, lettuce, melon, mulberry, mustard, oats, olive, onion, pea, pear, plum, pomegranate, quince, radish, rye, spinach, turnip, wheat
Mediterranean	Artichoke, asparagus, cabbage, cauliflower, fig, horseradish, parsley, parsnip
Southeast Asia	Banana, breadfruit, millet, orange, peach, persimmon, rice, soybean, sugarcane, tea, yam
Central or South America	Avocado, beans, cassava, corn, cranberry, gourd, kidney and lima bean, manioc, peanut, pineapple, potato, pumpkin, squash, sweet potato, tomato

[a] Adapted from Duckworth (1966).

domestication is debatable. The sweet potato or yam, *Dioscorea,* was brought under cultivation independently in South America, West Africa, and Southeast Asia (Coursey, 1975).

Domestication of fruits and plants used for their unique flavor also occurred during the Neolithic period (Zeuner, 1963; Zohary and Spiegel-Roy, 1975). The most significant early horticultural plants domesticated were dates, figs, grapes, and olives—the first three because they have a high sugar content, particularly when dried, and the last as a source of edible oil. In addition to onions and garlic, a wide variety of spices was used for flavoring: sage, thyme, fennel, wormwood, and others.

Food Preparation. Neolithic food preparation was primarily a home industry. A hypothetical chronology for the development of cooking techniques is given is Fig. 3. All of these originated in the home kitchen. Among the new food preparation techniques developed were sieving, salting, seasoning, pressing, alcoholic fermentation, acetification (vinegar formation), and bread making. It is interesting to note that some of these are still used in the home, while others are employed by commercial food processors.

The making of fermented beverages from plant saps predates the domestication of grapes, according to Forni (1975). Birch and palm wine were produced and consumed immediately by Paleolithic cultures. Preservation

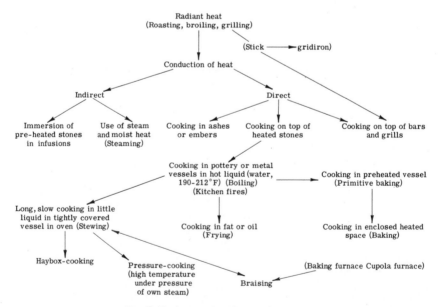

Fig. 3. Evolution of cooking techniques.

of alcoholic beverages depended on invention of containers (ceramic, wood, etc.). As communities became more settled, wine as we know it today became practical.

Fermentation as a method of food preparation has many varied applications, not just for producing beer and wine. In the Pacific (particularly in Polynesia), breadfruit has been fermented since the earliest settlements. Bananas, coconuts, and other plants were added to the breadfruit. A semianaerobic fermentation was developed, using pits in the ground under a sealing of leaves and stones (Yen, 1975). Modern fermentation industries are discussed in Chapters 7 and 8.

Salt is not only a taste additive and a food preservative but also has had ritual importance (Kare *et al.,* 1980). However, it is not universally added to food. Social custom seems to have been the main determinant of its use; once people became conditioned to using salt, they then stubbornly clung to its use. However, its ritual importance was well established: salt taboos surrounded menstruation, pregnancy, puberty, initiation rites, baptism, warfare, and so on. Symbolic meanings for salt are deeply ingrained in our language: in the expression, "You are the salt of the earth," to seal covenants, as a gift of welcome, to signal long life and health to a bride and groom, on sacrificial foods, and so on. It is well to remember that excessive use of salt is hazardous to health, particularly for people with hypertension, and that low-salt foods and diets are advisable for some, with a physician's recommendation (see Chapters 5 and 6).

Copper, Bronze, and Iron Ages

No specific dates mark the end of the Neolithic period and the successive development of the Copper, Bronze, and Iron Ages. In general, the period from 3500 B.C. to 1500 B.C. covers the most important developments. During this time humans learned to harness the wind, invented the wheel and the sailing boat, smelted ores, and began to develop an accurate calendar and a sophisticated written language. The city-state kingdoms with their privileged classes of nobility and priests were other features of this period. The domestication of plants and animals continued, with the new feature of conscious selection. Hoe culture, permanent fields, sophisticated irrigation, primitive plowing, and rotary millstones all appeared.

Food Preparation. Food preparation became more complex. Baking provided a variety of types of bread and confections. Leavened bread appeared for the first time. Lactic acid fermentation (pickling) dates from this period. Filtration, flotation (to separate olive oil), clarification (of beer and wine), and more sophisticated pressing methods were developed. Frying as a distinct type of food preparation appeared. Addition of fat and spices to meat to prepare sausages created a new type of food that could be stored (especially if smoked and dried). Dried fruits and fish were common. Cooling was practiced (with snow or by evaporation of water). In the city-states, the palace–temple economy demanded large-scale storage and preparation of food. (Note that these home or village industries are now carried out predominantly by commercial food processors, sometimes thousands of miles from the major consumer market.)

Perhaps the outstanding development of this period was the short- and long-range transport of food. In the cities the marketplaces attracted local farmers. The wheeled cart and the sailing ship led to long-range transportation of food. Olive oil from Crete and dried fish from Asia Minor were shipped far up the Nile.

Ancient Times to the Present

Middle East and Egypt

The earliest artifacts of the Neolithic period appear in the region from Asia Minor to Afghanistan. At a very early date there was considerable traffic between this region and Egypt, so it is not always possible to determine exactly where development of a new crop or food process took place.

In the Fertile Crescent area, in what is now Iraq and Iran, great civilizations began to develop by 3000 B.C. One outstanding feature was their dependence on beer. In the Sumerian temple economy, workmen received a liter of beer per day, low officials, 2 liters, higher officials, 3, and the nobility, 5! Labor accounted for 60% of the cost of producing beer, and as much as 40% of the cereals produced were converted to beer.

At first beer was made in the home, but later so great was the demand that an artisan class developed to produce it. The judicial code of Hammurabi (1728–1638 B.C.) had a special section on beer that prohibited sale at too low an alcoholic content (to prevent watering) or at too high a price—an early example of government control of the food industry. Eight types of beer from barley, eight from emmer wheat, and three mixed types are reported for the Sumerian period. Beer was not only highly nutritious because of its carbohydrate and alcohol content, but it was high in B-complex vitamins and it brought sensory and physiological pleasure to the consumer.

The main carbohydrate of cereals is starch, which has no flavor and which must be hydrolyzed to produce sweet-tasting and fermentable sugar. Malting was early discovered as a means of accomplishing this breakdown of starch. The grain is moistened and allowed to germinate. By the time it reaches the one-leaf stage a very active enzyme system develops in the seed by which starch is hydrolyzed to maltose and glucose. If the grain is dried at this stage the germination process is arrested but the enzymes remain active. The ground-up dried grain is called malt and is a source of enzymes as well as of sugar.

At first the malt was added to cereal to give it sweetness. Later it was used in beer making because it provided easily fermentable sugar to the yeast— thus speeding up the fermentation—and also because the enzymes speeded up starch hydrolysis of unmalted grain. Special beer utensils were used, probably because they had a large flora of yeasts. Beer was often flavored with lupine, skirret, rue, mandrake, wormwood, and other herbs or spices. The early Egyptian beer and wine makers distinguished the initial violent alcoholic fermentation from the slower secondary acetic acid fermentation that leads to vinegar formation. In a crude way they controlled the latter by stoppering the storage jars. The present-day counterpart is the commercial brewery with its huge fermentors, temperature control, selected strains of yeast, commercial malt, the use of hops, and proper packaging.

The hieroglyphics in the tombs of ancient Egypt provide us with a clear picture of food production and processing in the Bronze Age. Drying was practiced in a planned manner. Among the special achievements of the Egyptians were leavened (yeast-raised) and 40 other types of bread and cakes (Figs. 1, 4), sieving of cereals, and alcoholic fermentation.

The diet of the ruling and the rich classes was, as usual, more varied and abundant than that of the poor. Darby et al. (1977) quotes a bill of fare for

Fig. 4. Egyptian brewery showing grinding, kneading, mixing of yeasty residue from previous baking, rising of loaves of dough (rear), and man mixing water with bread with his feet. Thebes about 2000 B.C. (By H. M. Herget, © National Geographic Society; from Hayes, 1941).

King Unas of the Sixth Dynasty in Egypt (ca. 2600 B.C.): "milk, three kinds of beer, five kinds of wine, ten kinds of loaves, four of bread, ten of cakes, fruit cakes, four meats, different cuts, joints, roast, spleen, limb, breast, tail, goose, pigeon, figs, ten other fruit, three kinds of corn [i.e., wheat], barley, spelt, five kinds of oil, and fresh plants." A considerable degree of sophisticated food production and culinary art is obvious from this list.

Overeating and obesity, overdrinking and drunkenness were well-known problems among members of the affluent class in Egypt, as elsewhere in the ancient world.

Wine making first appeared in the Jemdet Nasr period in Mesopotamia. Wines were produced in Egypt in 3000 B.C. Possibly wine making reached its first peak there, although the climate was really too warm for grapes for wine production. The vineyards were developed as a part of the gardener's tasks, largely from arbors. Grapes were crushed by foot, and bag-type presses were used to separate the skins and seeds from the fermenting wine. Wines were expensive and were first used for ritual purposes by the rulers and priests. Several types of wine were recognized, such as Antyllon and Bueto. Date, palm, fig, and raisin wines were also produced (Darby et. al., 1977). The danger to workers in unventilated quarters of the carbon dioxide produced during fermentation was recognized, and some primitive filtration was practiced to separate the wine from yeasts. Egypt not only made wine but imported it from Greece and Asia Minor. Egypt also borrowed foods and processes for their preparation from Asia Minor and central Africa (Darby et al., 1977).

Lactic acid fermentation was known very early in Egypt. This fermentation was accomplished by using lactic acid bacteria in high-salt brines. Many pickled vegetables were thus prepared. Acetic acid fermentation (whether accidental or planned) was known in Egypt. Flavored vinegars were used as a condiment and also to preserve vegetables and meat.

Spices have occupied an ancient and peculiar position in the food industry. They yield few calories or vitamins; yet they stimulate the appetite and hence increase food consumption, thereby indirectly improving the body's nutrition. (Note: to say spice "aroma" is redundant. The Greek word for aroma means spice or spice material!) Not only do spices add a pleasant odor but also some have acceptable taste and pain sensations. Spices appear to have been used in early times to improve the sensory quality of foods or to cover up undesirable sensory properties.

Greek Period

The Greeks used a wide range of foods inherited from the East and South (Vickery, 1936). To these they added olive oil and seafoods such as mussels, octopus, and oysters. (Other early civilizations also gathered seaside fish and shellfish for food.) Olive oil production was especially important in Crete, possibly as early as 1500 B.C. (and even earlier in Asia Minor). To separate the oil, flotation was developed as a food process. Special crushers that did not break the seeds were developed (Fig. 5). Special presses were used to remove the residual oil from the crushed pulp. The best oil was produced before the olives were overripe.

Oil was used as a food, in cooking, in religious rites, on the body, and as a preservative for foods and beverages, to exclude air. Most important, it was exported to Greek colonies, to Egypt, and later to the Roman world. Wine was also an item of export from Greece. Greek amphoras (clay vessels of about 3 gallons capacity) from this era have been found throughout the Black Sea and Mediterranean regions, and as far up the Nile River as the Sudan. The amphoras were often lined with pitch and probably tasted of turpentine or petroleum. Herbs and spices were often added. Greek wine merchants dipped a sponge into the wine and smelled it when squeezed as a primitive form of sensory analysis. Nearly 100 different types of Greek wines have been identified. The amphoras were often identified with the place of production and the wine maker as an early form of a standard of identity or quality.

Still, the wines from this period cannot have been very good. They were almost always drunk diluted with water. This reduced the vinegary and lactic acid flavor and the alcohol content. Excluding air from processed foods and beverages did not progress very far until Louis Pasteur offered an explanation of the dangers of aerobic microorganisms in food spoilage.

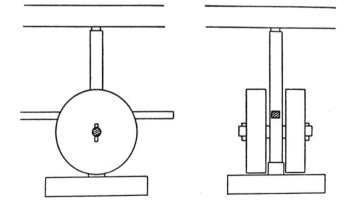

Fig. 5. Reconstruction of a crusher described by Columella for olives. The fruit is spread on the flat circular trough and the millstones are turned around the vertical pivot. Note clearance between millstones and trough, which prevents the olive pits from being crushed. The horizontal spokes are about waist high. First century A.D. (From Singer et al., 1954–1958.)

The Greeks developed new methods of cooking: special grills, skewers, sauces, and others. Elaborate feasts were given in the fifth century B.C. The meal was more organized as to time of day and dishes than in earlier periods.

Roman Period

Large-scale trade in foods over both short and long distances, characterized the Roman period. One of the reasons for the expansion of the Roman Empire was to obtain more food for Italy. Egypt, Spain, North Africa, and England were all sources of wheat. There were also many improvements in agriculture: fallowing, fertilizing, use of legumes in rotation, and improved threshing, among others.

Large-scale milling was a special feature of the Roman period. In the pushing mill both stones were flat, the upper one having a hopper with a slit in it so that the grinding surfaces were continually supplied with grain. Large ovens (Fig. 6) and mechanical kneading, instead of feet and hands, were used. The Roman millers were organized into guilds. In the late Roman Empire they were civil servants, as bread was the principal form of government dole.

One of Rome's greatest contributions was the spread of Roman crops and tastes for foods throughout the empire. The Romans encouraged the culture of grapes throughout France, southern Germany, Austria, Hungary, and Romania. Wine was an important item of trade. The Romans introduced fining (a method of clarifying wines), heating (a form of pasteurization), and boiling down of grape juice. The wooden barrel appears later, possibly as an idea of the Gauls.

Pressing, one of the most troublesome operations of food processing, was greatly improved during the Roman period. Mechanized beam and lever-and-screw presses (Fig. 7) were developed. Finally, the Romans developed

Fig. 6. Roman baker's oven. The men on the right are kneading dough. Part of a frieze from the monument of the baker Eurysaces. Rome, first century B.C. (From Singer *et al.*, 1954–1958.)

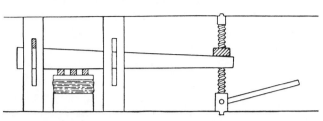

Fig. 7. Pliny's first lever-and-screw press. The screw has either one undercut bearing in the floor, or, as here, bearings in floor and roof. There is a second pair of slotted posts on the right of the press-bed. For filling, a beam is pushed through the slots in these posts and below the press-beam. The screw is then turned to bring down the end of the press-beam. First century A.D. (From Singer *et al.*, 1954–1958.)

a wide variety of new foods and methods of food preparation. Apicius' Roman cookbook (Guégan, 1933) is an example of how sophisticated food processing had become by then. Adulteration was also common (Vohling, 1936), e.g., rose wine made without roses, and spoiled honey treated to make it salable.

Far East

Needham (1954–1962), Arnott (1975), and Chang (1977) emphasize that the importance of Chinese developments in agriculture and food preparation has not been recognized in the West. Steaming—a method still used in Chinese cooking—developed there in the Neolithic period or earlier.

The Shang dynasty, about 1600 B.C., left writings on bone from which we have learned much about Chinese foods and food processes.

Rice had been first cultivated earlier (about 5000 B.C., according to Arnott, 1975). Wheat spread to China from the Middle East. Milk and milk products were rare.

By 1000 B.C. ice was used for refrigeration. Before 225 B.C. the animal-drawn plow, terracing, and irrigation had greatly increased food production. Millet was produced in China before the Christian era. The well-documented trip of Chhein to the Middle East in 138 B.C. resulted in the introduction of grapes, alfalfa, and new breeds of horses into the Far East. Later chives, coriander, cucumbers, pomegranates, sesame, safflower, and walnuts were brought to China from the West. China sent the orange, peach, and pear westward.

By 300 A.D., tea and water-run mills were common. Marco Polo observed a highly sophisticated cuisine and agriculture in the Far East in the thirteenth century. Before the nineteenth century the West got the idea of rotary fans and winnowing machines from China. Distillation, however, was not a Chinese invention (Needham, 1954–1962, Vol. I).

The civilization of India is equally ancient, and for religious reasons the diet was largely vegetarian. However, the early Aryans in the area (about 2000 B.C.) ate meat. Humped-back cattle, buffalo, goats, domestic fowl, sheep, pigs, camels, and elephants, but not horses, were domesticated in early India. Cotton originated there. Ghee (clarified butter fat), mixed spices (curries), and rice were used very early. The caste system introduced limitations on general food preparation. Soma appears to have been an alcoholic beverage, probably made from honey; it probably was developed before beer or wine.

Africa

The contributions of Egyptian civilization have already been mentioned. A number of plants from central Africa were diffused outside the continent: for example, sorghum and some related plants spread to the Far East by 2000 to 1500 B.C. (Arnott, 1975).

Middle Ages

The invasion of the Mongols introduced buckwheat to the West. The Crusades resulted in the importation into western Europe of a number of new varieties of fruits and vegetables from the Middle East. Various types of pastas developed in Italy; possibly Marco Polo introduced the idea of pasta from China. Kitchen stoves were invented, freeing the cook of dependence on the smoky and inconvenient fireplace. Water mills increased throughout Europe—there were 5624 in England in 1806. Trade in local products of special quality or need continued—dried fruits, spices, and wines from the Mediterranean area to northern Europe. Spanish olive oil, lard, and hams were widely traded.

The introduction of the iron plow (sixth to seventh century), the horseshoe (ninth century), and the horsecollar (soon after) greatly increased agricultural production, especially in Germany. The threefold rotation system (one third of the land remained fallow, summer crops were on another third, and winter crops on the remaining third) was a great improvement.

Distillation began to be used in Italy about 1100 A.D and was common throughout Europe from the thirteenth century (Figs. 8 and 9). Liqueurs (from fruits or herbs and spices) were produced in the fifteenth century. Distilled alcoholic beverages—brandy, gin, and rosoglio (alcohol and raisins)—were much favored during the plague (1348–1352). By 1360 taxes were increased in Germany to prevent excessive consumption of schnapps (a kind of vodka).

Fig. 8. Distilling apparatus. Brick-built still in which the alembic head is cooled discontinuously by a water trough. (From Estienne, 1567.)

Although trade was restricted, especially in certain periods, it was by no means nil. Besides the trade from the Mediterranean to the North, there were spices from Constantinople, wheat, garlic, onions, and onion seed from England and France to the South and East, rye from Poland (from the fourteenth century), apples from Normandy, Bordeaux wines from France to England, "white" cured stockfish from Norway, herring from Holland, and so on. Wines, beers, and cheese were very widely traded. By the sixteenth century trade in foods had expanded throughout much of the world.

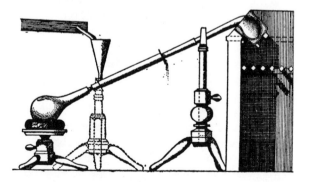

Fig. 9. Von Weigel's still with countercurrent cooling (1773). The hot distillate flows down from the retort on the right to the receiver, while the cold water flows upward through the water-jacket. (From Singer *et al.*, 1954–1958.)

Great Britain

The history of food production and processing in the British Isles from the thirteenth to the twentieth century is important because English methods of preparing food were so dominant in the colonial period of America. Similar developments, however, took place in other countries. The tragedy of this period is the richness of the diet of the rich, the nobility, and the guilds, and the poverty of the diet of the poor. Drummond and Wilbraham (1958) Burnett (1966), and Henisch (1976) have emphasized this fact. It is even more tragic to contemplate that, despite the marked improvement in the diet of the poor during the latter part of this period, there were still widespread dietary deficiencies in the nineteenth century in a country so rich as Great Britain.

Hunting continued to supply a variety of meat throughout this period—particularly for the wealthy landowner. The English garden was (and still is) rich in vegetables and herbs. Herbs and imported spices were widely used. (Edward I, 1239–1307, spent £1600 on spices in a single year!) Bread-making guilds were organized in the eleventh century, and, along with the Assize of Bread in 1266 and the Bread Act of 1822, they provided a fair measure of nutritious bread.

The reign of Elizabeth I featured (for the wealthy landowners, rich merchants, and nobility) massive dinners with dozens of dishes. These were consumed course after course but with much choice for each diner. At this time England imported 56 French wines and 30 more from Italy, Greece, and Spain. The Elizabethans loved sweets, even adding sugar to their wines. Nevertheless, beef, bread, and beer (or ale) were the staples of the Elizabethan diet for most of the population. There was widespread overconsumption of distilled alcoholic beverages, particularly gin.

In Stuart England, gorging was even worse. Even an ordinary dinner for the upper classes might consist of as many as 30 dishes. During this period Portuguese wines became popular. Consumption of coffee, tea, and chocolate also increased. Bananas and pineapples were imported in increasingly larger quantities.

From the thirteenth to the sixteenth centuries, eating meat on fish days or during Lent was a serious offense. Elizabeth I had two fish days per week, which Drummond and Wilbraham (1958) suggest was due to the government's desire to encourage shipbuilding. Henisch (1976) notes that the number of meals was reduced during Lent and that cooks devised numerous ways of preparing fish. Fasts were traditionally followed by feasts.

By the late seventeenth century the British diet had improved. The potato was popular and garden vegetables were in favor. Even the sick had special diets. St. Bartholomew's Hospital in London changed its menu daily in 1687. The old meat-fish order of serving was changed to the modern soup-fish-meat order in the seventeenth century.

Factors Influencing Food Supply and Processing

Food Adulteration

The protection of the consumer against dishonest food merchants has occupied the attention of civil authorities from the earliest times. At Constantinople in the late Roman Empire, stringent measures to prevent adulteration of spices were adopted. Between the thirteenth and sixteenth centuries all of the major branches of the food industry became subject to detailed regulations designed to protect the consumer. Foods under regulation included wine, bread, ginger, saffron, pepper, meat, beer, and ale. There were problems of short weight, inferior raw products, deterioration, and adulteration. The scientific developments and legal constraints of the nineteenth and twentieth centuries have brought most of these practices under control (see Chapter 9).

Great Britain. London had a meat-market overseer in the fourteenth century. In 1319 he succeeded in putting a butcher in pillory for selling putrid beef. The Company of Grocers were (and are) the keepers of the Great Beam (standard weight). They tried to prevent short weight and even supervised removal of impurities. Henry VIII forbade use of hops in beer, and "aleconners" (tasters) reported on dilution or illegal flavoring of beer. Early spice inspectors were called *garbellers*. They relied on appearance, taste, and smell to detect spoilage and adulteration. Artificial wine was detected in London in 1419. Wines were "sophisticated" (adulterated) with starch, gums, sugar, blackberry juice, elderberries, tournesol, and other substances. These practices may have injured the wine's quality but did not constitute a serious public health hazard.

Because alcoholic beverages were also heavily taxed, the question of their adulteration was important. If gunpowder into which the distilled spirit was poured would ignite, there was enough alcohol in it to call it "proof" spirit. This turned out to be about 50% alcohol. The famous English chemist Robert Boyle developed a hydrometer in 1675 specifically for alcohol determination. Later, special hydrometers were developed for analysis of a variety of food products.

After the eighteenth century food adulteration was sometimes dangerous. Vinegar was adulterated with sulfuric acid, green vegetables with copper (to improve color), beer with red pepper, tobacco, or licorice, and wine with sugar of lead (lead acetate)!

Pasteur's studies on the microbial origin of disease called the attention of food producers to the use of antiseptic agents. Some of these, if used in moderation, are not dangerous. However, in the last half of the nineteenth century a number of dangerous agents were used in foods: boric acid and borates in cream, formaldehyde in milk, and salicylic acid and benzoic acids in wines.

Accum's (1820) book is a landmark in this field. He called attention to the prevalence of food adulteration in Great Britain and gave chemical methods for their detection. (The microscope was also first used in detecting food adulteration during this period.) The *Lancet* (the leading British medical journal) substantiated and added to these charges. The result was the Food and Drug Act of 1860 and the Food Act of 1875. These and similar laws in other countries have gradually brought the more dangerous and fraudulent practices under control. (See Chapter 4 for a discussion of food-safety processing and Chapter 9 for a survey of food laws and regulations.)

One curious footnote is the history of a food adulterant that produces ergotism, a toxic condition caused by eating rye bread prepared from grain infected with the fungus *Claviceps purpurea*. It was noted as early as 1582 and was called St. Anthony's fire because people who ate infected bread were said to be "devoured by an invisible fire."

Famine

The history of man is one of a nearly constant struggle for food. Lack of food is due to the failure of agriculture to produce food, of processors to process and preserve it, or of industry to transport it from regions of surplus to regions of deficit. Failure to produce sufficient food has been due to unfavorable climatic conditions such as drought, excessive rainfall, frost, and winter killing.

Egypt. Famines occurred under local conditions in the ancient world. The familiar biblical story of Joseph indicates that the ancient Egyptians stored food in silos for as long as 7 years to avoid periods of low agricultural production due to failure of the Nile floods. Darby et al. (1977) noted various vitamin deficiencies found in Egyptian mummies but stated that nonetheless the poor generally had an adequate diet, even in times of drought, owing to governmental rationing. The poor were also relatively immune from the taboos observed by the clergy and nobility that affected such foods as certain species of fish, pork, beans, and onions. There were periods when beef was tabooed, according to Darby et al. (1977). These authors attribute the avoidance of pork to a desire on the part of the Israelites to set themselves apart from their Egyptian masters, or as a rejection stemming from the association between swine and false gods of agriculture. Alternately, perhaps it was intended to differentiate the Israelites from the pig worshipers.

Middle Ages. In the Middle Ages famines were usually local. In 1560, 1577, 1587, and 1596, however, there was widespread starvation in Great Britain even though the British were better supplied with food than the Europeans. As the population increased, famines in Europe affected rather wide areas.

The discovery of the Americas, which introduced new foods to Europe, and the advent of the railroad and steamship, permitting rapid transport over long distances, gradually reduced the incidence of famines. Nevertheless, as late as 1943 there was a general famine in the Bengal region of India, and since then there have been several localized famines in Africa and Asia.

Current estimates of the number of people suffering from hunger and malnutrition vary. The estimate given by the American Chemical Society (ACS) (1980) of 450 million of a world population of 4 billion is as good as any. The ACS emphasizes that although most of the hungry and malnourished are in the developing countries, there are substantial numbers of people with dietary problems in the developed nations, such as the United States (e.g., eating too much or too little, or a diet of poor quality). The basic problem, which is becoming more difficult to solve, is to produce enough food and to distribute it to the people who need it. In addition, losses must be minimized during processing, storage, and distribution. (For further discussions on food supply see Chapter 2; on nutritional diseases, see Chapter 6.)

Growth of the Commercial Food Processing Industry

With the invention of modern printing methods, both new books and translations of books by Roman writers appeared, spreading knowledge of old and new methods of food production and processing. Most of the essential aspects of food production were known before the Christian era (Table 1). What developed during the Middle Ages was commercial food processing and preservation, a trend that continues today and has resulted in less and less home food preparation. The development of mechanized large-scale operations has accelerated the trend. The Industrial Revolution contributed an increasing number of power sources, which have been gradually applied to food processing and preservation.

Consider the milk and cheese industry. It remained a cottage or semicottage industry until the mid-ninteenth century. Since then, cream separators, pasteurization, pure bacterial cultures, and a variety of mechanized equipment operations have enabled it to become almost completely automated.

Beer production was no longer a cottage industry even in the eighteenth century, although Oxford colleges continued to brew their own beer until the twentieth century. Gradually the breweries became larger and the equipment more mechanized. Finally, in the 1960s, continuous in-line commercial beer production was achieved.

Recently the food industry has not only produced processed foods but has been marketing foods prepared for cooking or partially or wholly cooked. Freeze-dried coffee, boil-in-bag vegetable items, TV dinners, and pre- or partially grilled hamburgers are included in these recent commercial developments. The big advantage of industrial food production is the reduction in cost and the better quality control as compared to cottage production. Modern commercial food processing operations are extremely complex, requiring sophisticated biological, physical, and chemical controls and highly trained personnel. Computer-assisted systems are now widely used for both monitoring and controlling operations (see Chapter 8).

Impact of War

Soldiers must be fed if they are to continue to fight. Foraging reduces their effectiveness as fighters. When the war is carried on at some distance from the home base this imposes special logistic problems. The food should be light in weight, easily packed and carried, nutritious, not easily spoiled, and appealing. In general, it is better to send foods that can be easily prepared at their destination. Wars thus impose special problems on the food industry. Large-scale production may have to precede actual combat by several years. Safe storage of the food is necessary. Food trains or ship convoys to transport the food have to be arranged.

Bread or grain for bread was the mainstay of all early armies (Ashley, 1928; Jacob, 1944). Dates, raisins, nuts, and other high-calorie foods were also popular. Dried meat and fish were widely used. Pemmican (dried meat pounded with melted fat and berries) was favored by the American Indians. The Boers of South Africa ate biltong (dried salted meat) as an item of army issue. Jerked beef or charqui and Hamburg beef were similar products.

There are many specific examples of the effects of wars on food processing. The French government, after the Revolution of 1789 offered a 12,000-franc prize for a new method of preserving food in a stable and nutritious condition. The Napoleonic Wars stimulated the search. As early as 1795 Nicolas Appert, a brewer and later a confectioner and cook, successfully preserved various foods in sealed jars heated in boiling water; he was awarded a prize for his discovery in 1810. The U.S. Civil War greatly expanded the canning industry in this country—about sixfold. The army bought canned California fruit for officers' messes. Canned goods of all types were a general item of issue, especially in the Northern army.

The British blockade during the Napoleonic Wars cut France off from her normal West Indian sugar supply. To meet the need, Napoleon asked the French botanists to improve the sugar content of sugar beets. They were so successful that Europe eventually became nearly independent of the West Indies for sugar.

In World War I the Allied blockade forced the Germans to develop ersatz foods (coffee, cooking oils, etc.). At the same time the success of the German U-boats forced Great Britain to add 10% cornmeal to wheat flour for bread.

World War II resulted in rationing, fortification of foods, and use of substitutes in Great Britain. Concentrated orange juice was imported for its vitamin C content. Rose hips, which are high in vitamin C, were collected in Scotland and a high-vitamin C syrup was produced. Milk and cheese were in short supply, hence calcium intake was low. To supply calcium, chalk was added to bread ingredients. It is believed that due to the country's intelligent ration program, British civilians were better fed, from the nutritional point of view, at the end of the war than at the start (see Pyke, 1968, for a general discussion).

During World War II the production of dehydrated foods was greatly increased and improved, especially in the United States. Dehydrated fruits and vegetables had been produced during the Civil War but they were generally of poor quality. However, by the 1920s dried grapes (raisins), peaches, and apricots were regular items of commerce. In World War II dehydrated milk, eggs, onions, carrots, cabbage, potatoes, and other foods, most of usable quality, were produced in large quantities, both for the armed forces and for civilians.

Wars have not always had desirable effects on the food supply. Shortages caused by blockades have been especially serious for children and older people. (The siege of Leningrad in World War II is a pertinent example.) Food production often suffers in wartime for lack of labor, fertilizers, and other needed resources. In future wars (God forbid!) adequate food processing equipment, even spare parts, may not be available.

Influence of Religion

Religion also has influenced food production and processing by demands for certain kinds of foods or by prohibition of others. Unleavened bread is needed by orthodox Jews at certain times of the year. The high priest of Jupiter in Rome ate only unleavened bread. The Roman Catholic Council of Florence in 1409 specified only wheat bread and grape wine for the Holy Communion. Pork was and still is prohibited to both orthodox Jews and Muslims (p. 73). Alcohol is not used by orthodox Muslims. Beef was and still is not eaten by most Hindus. Not only do religions demand certain foods, but they may have set rules for their preparation in meticulous detail.

The former "fish-on-Friday" requirement of the Roman Catholic church profoundly influenced the diet of Catholics for centuries and at the same time was a great encouragement to the fishing industry! Fasting was a way of life in medieval society (Henisch, 1976). Rigid diets were prescribed as a penance; individuals justified it as moderation for the sake of health. The 6-weeks-fast during Lent was strictly enforced by both the Western and Eastern churches. (Today Muslims still observe a similar month of fasting, when food may be eaten only at specific times of day.) In medieval European society various subterfuges or substitutions occurred—for example, using baked, grilled, boiled, stuffed, and fried fish with various spices and sauces, or wild birds. The prohibition on butter, eggs, and meat dishes posed a difficult task for the cook. Alcoholic beverages were generally exempt. Manual laborers, the sick, and the poor also were given a dispensation for eating meat and other forbidden foods during Lent. (For further information on food habits and taboos see Chapter 3.)

Exploration

Explorers and travelers have been bringing new varieties of animals, fruits, and vegetables to their home countries for many centuries. The great explorations of the fifteenth to seventeenth centuries greatly expanded this practice.

The discovery of the New World had an even more profound influence (*see* Fig. 2). Perhaps the most important new foods found in the Americas were the potato, corn (maize), and cassava (manioc, tapioca). The potato probably originated in the southern Peruvian or northern Bolivian highlands (Ugent, 1970). Although imported at an early date, it was not accepted as a food in Europe until after 1770. Thereafter it became a staple food in Germany, France, Great Britain, and Ireland. Corn was also accepted slowly but is now used thorughout the world as a food not only for people but also for poultry and livestock. There is some evidence for the pre-Columbian presence of corn in Africa and Asia (Jeffreys, 1975), but the main dispersion was almost certainly post-Columbian. The cassava was introduced into Africa by 1500 and into Asia shortly thereafter. In many areas it has become a major food crop (Moran, 1975). It is a product that requires special treatment to eliminate a toxic ingredient before use. The tomato and capsicum peppers were also adopted reluctantly in Europe. Now they constitute a significant part of the diet in Spain, southern France, Italy, Hungary, Thailand, and elsewhere. Other New World foods now used in Europe and elsewhere are the turkey, peanut, snap and lima beans, pumpkin, squash, pecan, cranberry, black walnut, pineapple, and cocoa.

The spread of New World foods to Europe was rapid and widespread. Less well known is the adoption of New World foods in tropical Asia. Burkill (1966) reported that Malay villagers grew and used the following New World foods: cashew, cassava, cherimoya, chilis, corn, custard apple, guava, passion fruit, peanut, pineapple, potato, pumpkin, sapodilla, soursap, star apple, sweet potato, sweetsap, and tomato.

Tea began to reach Europe from Southeast Asia early in the seventeenth century. The price had been so reduced by 1750 that tea became a popular beverage. Coffee orginated in Arabia, but the first supplies for Europe came from Ethiopia. By the seventeenth century it was so popular in Europe that the coffee plant was introduced into Java and Latin America to supply the demand.

Transportation

The revolution in transportation in the nineteenth century—primarily through the railroad and steamship—reduced the cost of transportation and greatly speeded up delivery of food. It also reduced losses in transport. Most importantly, it made perishable products easily available in the large cities. Until 1800 trade in fresh dairy produce was strictly local; by 1850 Irish butter was regularly available in London. Truck farming was greatly stimulated by the railroad. Out-of-season fruits and vegetables were supplied to distant markets. This is especially true now with air transport.

Mechanical refrigeration greatly increased the influence of transportation. Ice-making machines were patented in the first half of the ninteenth century. In 1877 the steamer *Frigorifique* successfully brought a boatload of refrigerated fresh meat from Buenos Aires to Rouen in 110 days. In 1886, 30,000 carcasses of mutton were shipped from the Falkland Islands to London. Australia and New Zealand have shipped huge quantities of meat to Europe since 1890. The first refrigerated railroad cars were in use in 1865. They have been particularly important in this country in transporting perishable meats, fruits, and vegetables long distances without spoilage. Nevertheless, as much as 40% of the fruit and vegetable products harvested in the United States spoils before it can be consumed (American Chemical Society, 1980).

Impact of Invention and Scientific Discovery

The Industrial Revolution

From the preceding discussion it is obvious how great was the impact of the Industrial Revolution on the food production and processing industry. The Industrial Revolution did not start suddenly. Ramelli's hand mill (Fig. 10) was made in 1588. It was quite elegant, with a rotating roller and spiral grooving. The central drum was offset slightly so that grain falling freely into the wider section was compressed and more easily ground. Watt's steam engine and Bessemer's process for making steel from pig iron (1856) greatly accelerated the Industrial Revolution.

The impact of the Industrial Revolution on food production was spectacular. Before 1700 most agricultural work was done by hand or at best with the

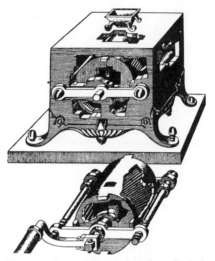

Fig. 10. Ramelli's portable iron roller mill for grinding flour. Both the roller and the interior of the drum are grooved; they are slightly tapered so that adjustment of the long screws alters the fineness of the grind. The grain is placed in the little hopper and the flour emerges from the spout (1588). (From Singer *et al.*, 1954–1958.)

help of draft animals. The seed drill was invented in 1700, and primitive threshing machines were made as early as 1780. McCormick's harvester appeared in 1834. Together with his seeder, it revolutionized grain production in this country and elsewhere. The combined threshing and cleaning machine dates from 1848. The development of the steam engine made these self-propelled and more efficient (Fig. 11).

Fig. 11. Threshing in California, 1883. The work was done on a large scale, requiring many men. In the figure, two "tablers" are continuously employed in feeding material into the "self-feeder." At the right of the machine another team is continuously employed hauling away the straw, part of which is used in the furnace of the steam engine. (From Anonymous, 1883.© 1883 by Scientific American, Inc. All rights reserved.)

Flour Milling. The milling of flour also changed. Roller mills were used in Hungary in 1840. Mechanical batting (sieving) of flour was introduced as early as 1500, but power-driven roller mills and cylindrical bolters did not follow until much later. Machines for mixing, preparing, rolling, and kneading the dough were developed in the nineteenth century. These were followed by automatic machines for dividing, weighing, and molding the loaves. The Perkins steam-heated ovens replaced the older coal- or coke-heated ovens, permitting closer control of the time and temperature of baking. Introduction of pure yeast cultures gave increased control of bread quality.

The cracker industry underwent similar changes. Machines for mixing and preparing the dough, rolling it to the proper thickness, and cutting it to the proper shapes and sizes (even to lettering or impressing patterns onto the crackers) were invented. By rigid control of the raw materials and the time and temperature of baking, a wide variety of crackers of uniform flavor could be produced throughout the year.

Fish. Other industries also felt the impact of the Industrial Revolution. The herring industry was partially industrialized as early as the twelfth century by the Dutch, who set up quality standards and were very successful in smoking fish on a large scale. It was not until the nineteenth century, however, that fans and heaters were introduced. The "Iron Chink" machine was developed for the salmon-canning industry in 1903. It cut off the head and tail, split the fish open, cleaned it, and put it in hot water, automatically adjusting its operation to the size of the salmon! Of course, refrigeration further changed the fishing industry to permit fishing in more distant waters. The use of ice to transport fresh fish started in Great Britain in 1786—interestingly, following a report that this was a common practice in China. Now frozen fish from distant countries is readily available in this country as well as in Europe. Canning fish is a major industry.

Sugar. Sugar from sugarcane was produced in the Near East and Egypt in the late Middle Ages. Large-scale production developed in the Madeira Islands and the West Indies in the seventeenth and eighteenth centuries (Fig. 12). The original Egyptian process expressed the juice between rollers, boiled it down, and cooled it to crystallize out raw brown sugar. This was improved by adding limewater and blood to the juice and filtering. In the nineteenth century evaporation was carried out in large containers, and charcoal and other decolorizing agents were used. Not only was a purer product produced, but machinery was developed to produce a variety of sugar products: powdered, confectionary, cubes, liquid sugar, and so forth. Now, whether produced from sugar beets or sugarcane, the process is nearly completely mechanized.

Fig. 12. West Indian sugar factory. Vertical roller mill crushing the canes (left); channel to the first boiler, where the juice was reduced and skimmed (center); boiler house (right). In the second boiler the sugar was purified with lime and egg white, in the third and fourth it was concentrated to permit crystallization (1694). (From Pomet, 1725.)

In general, the industrialization of food processing has greatly reduced waste and improved quality control, resulting in less wastage and spoilage and more uniform quality.

Modern Scientific and Engineering Developments

The Scientific Revolution had profound effects on food processing, and it still does. Napoleon III offered a prize for a substitute for butter; the prize was won by H. Mága-Mouriés, whose patent was granted in 1869. His discovery, margarine, is now a great competitor of butter. Fat hydrogenation was patented in 1902 and was in commercial use in Great Britain in 1909. Hydrogenated fat rapidly displaced lard for cooking. Pure yeast cultures for bread have already been mentioned. Pure cultures for beer production were introduced into the industry in the late nineteenth century and led to better and more uniform beers.

Laval's centrifugal cream separator was introduced in 1877, permitting great savings of space, labor, and efficiency in separating cream from milk. The Babcock butterfat test provided a sound basis for payment of the producer and helped standardize the fat content of milk. Cheese production with close temperature control, use of pure microbial cultures and rennet, determination of acidity, and other innovations, greatly improved the quality. An evaporated milk process was patented in 1835, and Gail Borden improved the technology with his vacuum process in 1853. His sweetened condensed milk, developed in 1860, was soon accepted as a food product of excellent quality. Later, unsweetened concentrated milk, heat-sterilized in cans, became popular. A process for drying milk was patented in Great Britain in 1855, but a high-quality product was not developed until nearly a century later. Various types of milk products and milk substitutes are now being produced in large quantities.

As already indicated, World War II resulted in improved dehydrated foods, which have since been further developed. Freeze-drying can produce meats that are easily stored and quickly reconstituted. The process of potato dehydration has also been markedly improved. Potatoes had been dried and made into a flour in Peru for centuries. The process is laborious, involving freezing and treading to squeeze the water out. The modern process is largely automated and produces a stable product of uniformly high quality.

Canning. The commercial development of Appert's canning process (p. 22) was one of the important scientific developments of the food industry. The first cans were handmade of steel coated with tin. Durand's early process produced 10 cans per day; modern can-making machines can produce 1000 or more per minute. A similar story could be told about glass bottles and jars for processed foods.

Successful canning was carried on in Great Britain by Donkin and Hall in 1812. The British Navy was an early user of their products. By 1831 canned foods were carried on ships as "medical comforts," and by 1847 canned beef was an item of regular issue. Kotzebue, the Russian explorer, used English canned foods for his voyage through the Northwest Passage in 1815. Commodore Matthew Perry carried canned goods on his 1819–1820 voyage toward the North Pole. Samples of meat and carrots carried on this trip were found to be good 100 years later. The early canneries in Great Britain were largely small-scale batch operations.

Deggett obtained a U.S. patent for canning lobster, salmon, pickles, jams, and sauces in 1815. Underwood started commercial operations in Boston in 1817, and Kensett in New York in 1819. A Boston cannery processed pickles, jellies and jam, quinces, currants, and cranberries in 1820. Large-scale canning of seafood in Maine dates from 1843. Tomatoes were canned for Lafayette College (Pennsylvania) students in 1847. Espy canned large quantities of cherries, plums, gooseberries, pears, peaches, strawberries, and vegetables in Philadelphia starting in 1855. Borden canned condensed milk in 1855.

During the Civil War his entire output was used by the army. Salmon was canned on the Columbia river in 1866 and in Alaska in 1872.

Until about 1860 sterilization required 5–6 hours at 212°F (100°C). By adding calcium chloride to the processing water, its temperature was raised to 240°F (115.5°C), so that the sterilizing time could be reduced. The vacuum (steam) retort, introduced in 1874, further reduced the time needed for sterilization. Mechanical handling (up to 1000 cans per minute), vacuum packing, new and more resistant linings and coatings, high-temperature pressure cooking and other improvements followed. (For further details on canning see Chapters 7 and 8.)

Today the latest developments in science and engineering soon find applications in food processing. The effect of the Scientific and Industrial Revolutions on the food industries is by no means at an end. In this country the U.S. Department of Agriculture, the state agricultural experiment stations, and the food industry have been the main vehicles for suggesting applications of new scientific and engineering concepts to food processing. Generally, however, it has been the food processing industry itself that has developed the new or improved methods and equipment. It is of interest that the agricultural division of the U.S. Patent Office was established in 1839 and became a separate department in 1862. The Agricultural Experiment Stations date from 1874. They have had many and profound effects on food production in this country.

REFERENCES

Accum, F. C. (1820). "Treatise on adulteration of food and culinary poisons, exhibiting the fraudulant sophistication of bread, beer, wine, spiritous liquors, tea, coffee, cream, confectionary, vinegar, mustard, pepper, cheese, olive oil, pickles, and other articles employed in domestic economy; and methods of detecting them." A. Small, Philadelphia, Pennsylvania.

American Chemical Society (1980). "Chemistry and the Food System." Am. Chem. Soc., Washington, D.C.

Anonymous (1883). Threshing in California. Sci. Am., Suppl. 16, 6370.

Arnott, M. L., ed. (1975). "Gastronomy. The Anthropology of Food and Food Habits." Mouton, The Hague.

Ashley, W. (1928). "The Bread of Our Forefathers; An Inquiry in Economic History." Oxford Univ. Press (Clarendon), London and New York.

Bökönyi, S. (1975). Effects of environment and cultural changes on prehistoric fauna assemblages. In "Gastronomy. The Anthropology of Food and Food Habits" (M. L. Arnott, ed.), pp. 3–12. Mouton, The Hague.

Brothwell, D., and Brothwell, P. (1969). "Food in Antiquity. A Survey of the Diet of Early Peoples." Praeger, New York.

Burkill, I. H. (1966). "A Dictionary of the Economic Products of the Malay Peninsula." Minist. Agric. Coop., Kuala Lumpur. (Orig. publ., 1935.)

Burnett, J. (1966). "Plenty and Want; A Social History of Diet in England from 1815 to the Present Day." Nelson, London.

Chang, K. C., ed. (1977). "Food in Chinese Culture: Anthropological and Historical Perspectives." Yale Univ. Press, New Haven, Connecticut.

Coursey, D. G. (1975). The origins and domestication of yams in Africa. In "Gastronomy. The Anthropology of Food and Food Habits" (M. L. Arnott, ed.), pp. 187–209. Mouton, The Hague.

Darby, W. J., Ghalioungui, P., and Grivetti, L. (1977). "Food: the Gift of Osiris," 2 vols. Academic Press, New York.

de Castro, J. (1952). "The Geography of Hunger." Little, Brown, Boston, Massachusetts.

Drummond, J. C., and Wilbraham, A. (1958). "The Englishman's Food; A History of Five Centuries of English Diet." Cape, London.

Duckworth, R. B. (1966). "Fruit and Vegetables." Pergamon, Oxford.

Estienne, C. (1567). "L'Agriculture et Maison Rustique," p. 171. Chez I. Du-Puys, Paris.

Forni, G. (1975). The origin of grape wine: a problem of historical-ecological anthropology. In "Gastronomy. The Anthropology of Food and Food Habits" (M. L. Arnott, ed.), pp. 67–68. Mouton, The Hague.

Guégan, B., transl. (1933). "Les Dix Livres de Cuisine d'Apicius." Bonnel, Paris.

Hayes, W. C. (1941). Daily life in ancient Egypt. Nat. Geogr. Mag. **8**, 419–514.

Heiser, C. B., Jr. (1981). "Seed to Civilization. The Story of Food," 2nd ed. Freeman, San Francisco, California.

Heizer, R. F., and Elasser, A. B. (1980). "The Natural World of the California Indians." Univ. of California Press, Berkeley.

Henisch, B. A. (1976). "Fast and Feast; Food in Medieval Society." Pennsylvania State Univ. Press, University Park.

Jacob, H. E. (1944). "Six Thousand Years of Bread. . . . " Doubleday, New York.

Jeffreys, M. D. W. (1975). Precolumbian Maize in the Old World: an examination of Portuguese sources. In "Gastronomy. The Anthropology of Food and Food Habits" (M. L. Arnott, ed.), pp. 23–66. Mouton, The Hague.

Kare, M. R., Fregly, M. J., and Bernard, R. A., eds. (1980). "Biological and Behavioral Aspects of Salt Intake." Academic Press, New York.

Moran, E. F. (1975). Food, development and man in the tropics. In "Gastronomy. The Anthropology of Food and Food Habits" (M. L. Arnott, ed.), pp. 156–169. Mouton, The Hague.

Needham, J. (1954–1962). "Science and Civilization in China," Vols. I–IV. Cambridge Univ. Press, London and New York.

Pomet, P. (1725). "A Compleat History of Druggs," 2nd ed. R. & J. Bonwicke & R. Wilkin, London.

Pyke, M. (1968). "Food and Society." Murray, London.

Reed, C. A., ed. (1977). "Origins of Agriculture." Aldine, Chicago, Illinois.

Saffiro, L. (1975). Monophagy in the European Upper Paleolithic. In "Gastronomy. The Anthropology of Food and Food Habits" (M. L. Arnott, ed.), pp. 79–88. Mouton, The Hague.

Sebrell, W. H., Jr., and Haggerty, J. J. (1968). "Food and Nutrition." Time, New York.

Singer, C., Holmyard, E. J., Hall, A. R., and Williams, T. I. (1954–1958). "A History of Technology," 5 vols. Oxford Univ. Press (Clarendon), London and New York.

Stanford, D., Bonnichsen, R., and Morlan, R. E. (1981). The Ginsberg experiment; modern and prehistoric evidence of a bone-flaking technology. *Science* **212**, 438–440.

Tannahill, R. (1973). "Food in History." Stein & Day, New York.

Ugent, D. (1970). The potato. *Science* **170**, 1161–1166.

Vickery, K. F. (1936). "Food in Early Greece." Univ. of Illinois Press, Urbana.

Vohling, J. D. (1936). "Apicius. Cookery and Dining in Imperial Rome." Torch Press, Cedar Rapids, Iowa.

Wendorf, F., Schied, R., El Hadidi, N., Close, A. E., Kobusiervicz, M., Wieckowska, H., Issawi, B., and Haas, H. (1979). Use of barley in the Egyptian Late Paleolithic. *Science* **205**, 1341–1347.

Yen, D. E. (1975). Indigenous food processing in Oceania. *In* "Gastronomy. The Anthropology of Food and Food Habits" (M. L. Arnott, ed.), p. 147–168. Mouton, The Hague.

Zeuner, F. E. (1963). "A History of Domesticated Animals." Hutchinson, London.

Zohary, D., and Spiegel-Roy, D. (1975). Beginnings of fruit growing in the Old World. *Science* **187**, 319–327.

Chapter 2

World and United States Food Situation

The problem of the quantity and quality of the food supply of the world is of immense proportions and of the highest concern to all nations, developed or developing. To provide information and to meet emergency and long-range needs, a number of organizations have been created.

Many U.S. governmental agencies are active in agricultural, health, food, and nutrition research. Their estimated annual expenditures are about $5 billion, of which an unknown but substantial amount concerns research applicable to world food problems. Foreign expenditures in these fields are not known but are probably much larger.

World Food Organizations

United Nations

Three organizations now function within the United Nations in this field. The Food and Agriculture Organization (FAO) of the United Nations grew out of an earlier organization in Rome. It gathers and publishes technical and economic information on food and agricultural developments throughout the world. Improvements of levels of nutrition and greater efficiency in food production and distribution are specific objectives of the organization. The

FAO assists its members in attaining these goals. There is a biennial conference to set specific goals and policies (see FAO, 1977, 1979a, b, 1980a). The FAO has sent numerous survey teams to many countries to study specific food and agricultural problems and to suggest methods of solving them.

The United Nation's Children's Fund (UNICEF) was organized to meet emergency food needs of children. The FAO has cooperated with UNICEF in providing technical advice on production and processing in appropriate places and, in some cases, supplying essential equipment. UNICEF received the Nobel Peace Prize in 1965.

The World Health Organization (WHO) traces its history to the International Sanitary Congress in Paris in 1851 and the Health Department of the League of Nations in the 1920s. The present organization was started in 1948 and has its headquarters at Geneva, Switzerland. There are six regional offices. The WHO cooperates with the FAO on nutritional problems. One of its primary concerns is nutrition as it affects health: WHO studies have focused attention on deficient diets, safety of food supplies, and sanitation in food processing. The WHO also maintains a close watch on the incidence and spread of disease through the world.

Other Organizations

The Pan American Health Organization (PAHO), located in Washington, D.C., has six regional zones. It identifies nutritional problems in the Western Hemisphere, trains personnel, cooperates in research activities, and gathers health and epidemiological information and statistsics.

The Institute of Nutrition for Central America and Panama (INCAP) performs similar functions for its local area. It has conducted research and provides advisory services on nutritional problems to its six member countries.

The Inter-American Institute of Agricultural Sciences is an agency of the Organization of American States (OAS). Its headquarters are at Turrialba, Costa Rica. It has projects on food production and food processing, holds training conferences, and is especially active in training agricultural extension workers for local areas.

The Agency for International Development (AID) is the agency of the U.S. government (under Public Law 480) designed to help developing countries achieve economic strength and momentum so they can provide a better life for their own people, using their knowledge and resources. Billions of dollars have been spent in many countries. The program has generally been successful. Successful AID programs are now being or have been terminated in Brazil, Greece, Israel, Mexico, Spain, Taiwan, Venezuela, and other countries.

Public Law 480 (P.L. 480) is the basic U.S. legislation covering food aid programs. There is also a sales program, partially financed by the Commodity Credit Corporation. Repayment varies from 20 to 40 years with low interest rates. Donations of food are also provided for and are administered by AID. P.L. 480 provides that at least 75% of the food be allocated to countries with a 1978 gross national product (GNP) of less than $625 that cannot meet their immediate food needs from domestic production or commercial imports. The priority of AID is to help meet the nutritional needs of vulnerable groups. Maternal/child-health programs, food for workers on public building projects, and preschool and primary-school feeding are emphasized. Blended and fortified foods have been supplied for nutritional programs. Funds accumulated from the sale of P.L. 480 commodities can be used for programs in agricultural and rural development, nutrition, health services, and population planning. Since 1980 AID has also sponsored research in American universities and at the National Academy of Sciences (NAS) on international food problems.

Other U.S. agencies involved in food, nutrition, and health on an international basis include the foreign agriculture group in the U.S. Department of Agriculture (USDA) and the Peace Corps. The latter has worked in a variety of practical ways to improve health, nutrition, education, and the food supply in more than 50 countries.

Private agencies have also been active in the international food field. The Rockefeller Foundation has worldwide interests in specific projects, such as food production in Mexico and rice production in the Orient. The Ford Foundation has supported a variety of food production and processing projects. Jointly these two foundations created the International Rice Research Institute, located near Manila in the Philippines. It has had great success in developing new higher-yielding rice varieties. The Meals for Millions/Freedom from Hunger Foundation has also been very active in recent years.

Worldwide Food Situation

The worldwide food situation is periodically evaluated by the FAO (1977, 1979a,b, 1980a). Per capita food production in the developed countries before World War II was nearly three times as great as in the developing countries. Since 1968 the level of agricultural production in the developing countries has been increasing, except in Africa (Fig. 13, Table 3). For a projected population of 6 billion in 2000, food production must increase 50% just to maintain present food levels.

During the 1970s (FAO, 1979b), considerable efforts were made in the developing countries to raise levels of food production. There was a substantial

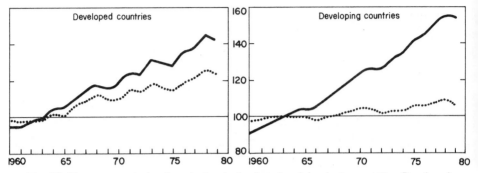

Fig. 13. Changes in agricultural production in developed and developing countries. Developed countries include the United States, Canada, Europe, U.S.S.R., Japan, Republic of South Africa, Australia, and New Zealand. Developing countries include South and Central America, Africa (except Republic of South Africa), and Asia (except Japan and Communist Asia). Solid line represents agricultural production; dotted lines, agricultural production per capita. (From USDA, 1980a.)

increase in the amount of irrigation and fertilization and also improved plant material (seeds, etc.). However, food production fell short of requirements, and the growing population compounds the problem. In 1979 the cereal-production gap in the developing countries rose to 85 million tons. Added to the burdens of the developing countries were increased prices and costs of transportation, especially petroleum products. The FAO report cited also indicates a continuing and long-term decline in the developing countries[1] share of world agricultural trade. This reduces their capital for importing needed foods.

TABLE 3

Index Numbers of per Capita Food Production[a,b]

	1968	1972	1976	1979
Europe (excluding U.S.S.R.)	100	102	108	114
U.S.S.R.	101	98	108	108
North and Central America	101	101	107	110
South America	98	98	104	107
Oceania	103	100	107	108
Asia	99	98	104	107
Africa	98	99	93	90
All above regions	101	99	103	105

[a]Data from FAO (1980b, p. 81).
[b]1969–1971 = 100. These index numbers were calculated by the FAO on a uniform basis employing regionally constant weights.

Food Security

The security of the world's food supply has remained precarious for several years. Carry-over stocks of food grains in 1979 represented only 18% of current consumption! The food crisis of 1973–1974 could be repeated at any time. In fact, as of the end of 1979 the FAO early warning system listed 26 developing countries as being affected or threatened by abnormal food short-ages as a result of poor crops, the effects of warfare, or difficult economic situations. The U.S. Department of Agriculture (USDA) (1980c) identified sub-Saharan Africa and South Asia as the critical food-deficient areas. Poor climate, political instability, and administrative inefficiencies contributed to the problem.

Production Patterns and Problems

Production by specific crops is given in Table 4. This shows clearly the relatively high yields per hectare in Europe and North America. Note, for example, that wheat production per unit area is three times as great in Europe as in Africa, whereas corn production per hectare in North and Central America is more than twice that in Asia. Observe how low potato production is in South America as compared to North and Central America. The low

TABLE 4
Yield of Major Crops[a,b]

	Europe[c]	North and Central America	South America	Asia	Africa	Oceania
Wheat	34.0	21.5	12.9	5.0	10.7	14.1
Rye	21.7	16.1	8.8	13.5	3.7	7.1
Barley	32.7	24.3	11.5	13.3	7.4	14.9
Oats	26.3	19.3	13.0	13.0	5.1	11.5
Corn	46.1	55.3	17.4	21.1	11.5	47.2
Millet	17.0	13.0	13.0	6.3	5.8	7.6
Sorghum	34.1	35.0	28.1	8.8	7.0	24.1
Rice, paddy	49.7	40.5	18.3	26.5	18.3	56.4
Potatoes	210.3	269.2	99.3	113.0	87.7	241.9
Sweet potatoes	103.8	72.5	100.7	84.9	64.6	53.8
Cassava	—	64.6	114.7	109.8	65.5	110.8

[a]Data from FAO (1980b).
[b]Results in 100 kg/hectare.
[c]Does not include U.S.S.R.

production per hectare, particularly in Central and South America, is related to absentee land ownership. Political changes have resulted in some shifts in land ownership. On the other hand, political unrest has interfered with food production in various countries. Whether it will permanently decrease production is not known.

The great wheat production is in Europe, North and South America, and Africa (Table 5). Production of corn in North and Central America is more than in all the other continents combined. On the other hand, Asia is the largest producer of rice, and Africa of cassava. Most of the world's potatoes are produced in Europe.

Technological Improvements. Recently, high-yielding strains of corn, rice, and wheat have been developed. Production per unit area has increased markedly through their use, particularly for rice in the Far East. There is no doubt that the new agricultural technologies (usually called the Green Revolution) now available could markedly improve food production in many parts of the world. Genetic technology offers further opportunities.

The food problem of developing countries would be less serious if the most modern technology to prevent spoilage and introduce better methods of food perservation were applied. Adoption of the new varieties and methods of culture and of improved processing technologies may be agonizingly slow. Their intensive use may also introduce new problems of pest and disease control. Reform of land policy and large-scale use of educational programs will be necessary. Some ticklish problems of priorities arise. (Is it better to increase production using the most modern insecticides and pesticides and risk some environmental pollution, or should one accept a lower production, food shortages, and even famine, and not use them?)

Soil Erosion. A major problem that is frequently overlooked is the loss of topsoil due to soil erosion. About 20% to 30% of the global cropland is losing soil at a rate that reduces crop productivity (Brown, 1981).

Calorie and Protein Inequities. Another way of looking at the inequality is on the basis of consumption, either by calories or by total protein or animal-protein intake. Data for 1975–1977 are given in Table 6. The table shows clearly the relatively high-calorie diet of Europe, North America, and Oceania and the low calorie content of the diet in the Far East and Africa and in some (but not all) South American countries. The FAO (1977) report shows that whereas the developed countries have 132% of their caloric requirement available, the developing countries have only 96%. Note also the very low total and animal-protein intakes in the Far East. However, there has been improvement in protein intake—though it is still far from adequate.

TABLE 5
Production of Major Crops, 1979[a, b]

	Europe	North and Central America	South America	Asia	Africa	Oceania
Wheat	83.9	78.4	123.4	135.3	8.9	16.4
Rye	11.5	1.1	0.2	2.7	—	—
Barley	68.0	17.2	1.0	32.1	3.8	4.0
Oats	13.9	10.8	0.8	1.7	0.2	1.5
Corn	56.3	213.9	29.0	62.3	23.9	0.4
Millet	—	8.6	0.3	21.2	9.6	—
Sorghum	0.7	25.2	7.4	23.0	9.8	1.1
Rice	1.9	8.1	12.4	345.5	8.7	0.7
Potatoes	121.9	19.5	15.8	36.9	4.6	1.1
Sweet potatoes	0.1	1.3	2.3	104.6	5.1	0.2
Cassava	—	0.1	29.9	41.0	45.0	—
Legumes	2.05	3.0	3.2	31.0	5.1	0.2

[a]Data from FAO (1980b).
[b]Results in millions of metric tons.

TABLE 6
Estimated per Capita Calorie and Protein Content of National Average Food Supplies[a,b]

Region and country	Calories/day	Total protein (gm/day)	Animal protein (gm/day)
World	2590	69	24
Europe	3410	96	53
U.S.S.R.	3443	103	51
North and Central America	3215	93	57
Asia	2276	58	12
Africa	2308	59	12
Oceania	3204	96	63
South America	2565	66	29

[a]Data from FAO (1980b).
[b]Unless otherwise noted, data are for years 1975–1977. In some cases tentative data are given.

Income Variations. Another factor influencing the world food problem is the lower income of people in the developing countries. Income is related to land policy. It is also influenced by the continuing growth of population, which is related to birth and death rates. It is a cruel fact that as the developing nations cut down their infant-mortality rates and increase their life expectancies, they also increase their populations and thus aggravate their food-shortage problems (Deevey, 1980).

Monoculture. Monoculture is a system of agriculture in which a single crop predominates. It almost always causes periodic economic sickness for a country and results in poor agricultural production and an unbalanced food supply. This leads all too frequently to malnutrition.

The developed countries, because of their greater per capita income, also require (or use) more food per capita. This reduces the amount of food available to the developing countries (Biswas and Biswas, 1979).

Projections

The FAO's most optimistic projections (1979a,b) (which assume a rapid improvement in economic growth rates of developing countries) show more adequate levels by 1985. If the FAO's pessimistic projections are used (these assume that little economic improvement will occur), then the number of hungry people will rise in the future to even more alarming proportions.

Inadequate Calories and Proteins. The more disquieting feature of the FAO's recent reports is that protein malnutrition, even with the most optimistic projections, will still be a serious problem in 1985. About one third of the world's population, especially infants and children, will then have an inadequate calorie/protein diet! If the weather is unfavorable or if there is a lack of fertilizers, seeds, or pesticides, serious crop failures and large famines could occur in several of the developing countries.

Impact of Population Growth. Even more disturbing is the tendency of population growth to "eat up" agricultural growth. Data for available food supply per capita in India are shown in Table 7. This shows decreases in per capita caloric intake in 1973 and 1974. During 1971–1974 India imported 1.2 million tons of total cereals (USDA, 1977). Thus, in spite of increasing home production and imports, calories available per capita declined! Manocha (1975) reported that food production doubled in India between 1950 and 1970, but on a per capita availability basis it increased only 18%. The average Indian still got only 80% of his daily caloric requirements in 1970!

TABLE 7
Food Supply per Capita per Day in India[a]

Year	Proteins (g)	Fats (g)	Calories
1972	50.3	28.9	2053
1973	47.0	27.3	1886
1974	48.0	28.4	1976

[a]From United Nations (1978).

The caloric, animal protein, and other protein content of the diets of various countries is shown in Fig. 14. Note the low levels of calories and animal protein in the diet of many countries.

In summary, if 2500 calories and 60 gm of protein per capita per day are considered a desirable food intake for an adult, more than two thirds of the present population is undernourished. Can we improve the dietary standards of the existing population if at the same time we continue adding approximately 70 million extra hungry mouths every year?

Solutions

Everyone agrees that we ought to try to help those who do not have sufficient food. Paddock and Paddock (1967), Borgstrom (1965), Ehrlich and Ehrlich (1970), and others predicted a global crisis in food supply by 1975. The Ehrlichs expressed grave doubt that the Green Revolution (increased food production through use of new varieties, new methods of production, use of fertilizers, etc.) would solve the food problem. Myrdal (1970) argued that even if the Green Revolution were successful it probably would not result in any significant reduction in malnutrition because of the increasing population. He also feared that lack of land reform and slowness to adopt labor-saving mechanization would reduce productivity. There is also the probability that, given an increase in food, those who are now under- and malnourished will increase their food needs and production will again be insufficient.

Priorities and Goals. The Paddocks' formula was for the United States to supply food primarily to nations that need extra food and can utilize it. But who is to decide which nations can best utilize food aid? Borgstrom (1965, 1980) was equally pessimistic. His solution called for a worldwide campaign for birth control and for increasing food production based on planning at a superstate level.

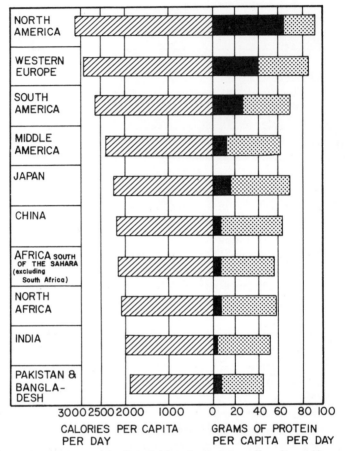

Fig. 14. A comparative picture of the diets of rich and poor nations. Key: diagonal lines, calories; solid area, animal protein; dotted areas, other protein. (From Manocha, 1975. Courtesy of Charles C. Thomas, Publisher, Springfield, Illinois.)

For a less pessimistic view of the world food problem (which has been specifically criticized by the Ehrlichs), see the Rockefeller Foundation symposium (Anonymous, 1968). The necessity of farmer education, increased use of fertilizers, and land reform are emphasized in this report. For a relatively cheerful view of the prospects to the year 2000, see Bennett (1963) and Brooks (1970). Even so, Brooks strongly recommends immediate technological improvements in the agriculture of developing countries: fertilizers, farm credit, foreign economic assistance, control of pollution, and so on. Moreover, he states, this technological revolution must be accompanied by population control so that by the year 2000 a stable population level is achieved.

Poleman (1975) also takes a more optimistic position. Population can be controlled (Singapore, Taiwan, Mauritius, and Sri Lanka are examples), increased incomes are essential, and Asia represents the major problem area. He does not believe that the Green Revolution has failed; rather, it has suffered "from lack of sustained attention." He envisages better methods of food production and substantial social and political changes. Failure in these areas will lead, he says, to civil strife on an "unprecedented" scale. Campbell (1979) also is optimistic if governments insure income and food distribution. He favors more mission-oriented agricultural research and national goals of self-sufficiency in the production of basic foods. He writes, "The jeremiads of doomsday men about the future of food production potential of the world are just plain wrong." The second Green Revolution may prove him right. There has been an encouraging decline in fertility in some developing countries (Anonymous, 1981). China has recently improved its food supply by increased production (and some importation). Harrison (1980) attributes this increase to the Green Revolution.

The general solution envisages improving the economies of the developing countries, hoping that as their GNP improves their population growth will slow down. Until this happens, more food must be found for those who are not producing enough to feed themselves. This is the ethical or religious and political view of the world food problem to which most of us subscribe.

As Borgstrom (1980) noted, "There are no simple solutions; if there were, we would long ago have implemented them."

Hayes (1981) reasons that the wrong thing to do is to give starving people (whose food supply is exhausted) more food—unless, he adds, it helps the nation to help itself. Because a number of developing countries are already populated beyond the carrying capacity of the country, Hardin (1974) argues that sending food leads to further destruction of the land and further impoverishment of the people, unless the country has resources to improve its carrying capacity. Simply supplying food will swell the population and thereby accelerate the disaster. To send food to a country beyond its carrying capacity is, he believes, wrong.

To many this is an almost obscene suggestion. It is certainly offensive to the accepted standards of decency of many people. Nevertheless, there is an inherent logic to Hardin's argument. In his view, the solution, will have to come by community self-sacrifice, birth control—which, he believes, will be

---

Carrying capacity is defined as the maximum number of a species that can live in an environment—indefinitely!—without degrading the quality of life for that species.

required in the future—being the most important. It is too soon to predict that abortion will be as common as he suggests, especially in countries where the culture is built around large families. If countries accept immigration they run the risk of overtaxing their carrying capacity. Yet who is wise enough to make such a decision?

In opposition to Hardin's premise that "more food means more babies," Murdock and Oaten (1975) do not believe that it is a question of "them or us." They estimate that proper expenditure of $10 billion per year by the United States (and a similar amount by the rest of the developed world) would measurably help solve the world food problem. The aid would include offering family-planning services, improving literacy, ensuring health care for mothers, providing grain, seeds, and fertilizers, and helping to increase irrigation. (See also Harrison, 1980).

Identifying the Malnourished. The FAO (1977) in its Fourth World Food Survey again emphasized the increasing population and the synergism between population and undernutrition and malnutrition.* The questions were raised, "Who are the malnourished?" and "How can they be identified?" Table 8 shows the severe protein-energy malnutrition in children in some selected poor land areas. A further chilling fact is that in a given country low-income families consume fewer calories per capita than do high-income families (Table 9). As the FAO study points out, this is true for the Malagasy Republic, Tunisia, and other countries in both rural and big-city populations.

According to the FAO (1977), the large-population countries with the most serious problem of inadequate food intake include Bangladesh, Brazil, Burma, Colombia, Ethiopia, India, Indonesia, Nigeria, Pakistan, the Philippines, Sudan, Tanzania, and Zaire. Smaller-population countries that have the same problem include Bolivia, Chad, El Salvador, Maldives, Mali, Mauritania, Niger, Peru, Somalia, and Upper Volta. Obviously, this list changes from year to year. Higher-income countries obviously have a wider choice of alternatives than do lower-income countries (International Food Policy Research Institute, 1977).

The agricultural import needs of the developing countries have also increased because of their growing industrialization; thus they must import more raw materials (cotton is a prime example). There is some evidence (FAO, 1979a) that some developed countries have become more protectionist since 1974–1975. National high-price support programs for locally produced goods in developed countries comprise one such nontariff barrier. It is difficult to see the justice in this situation.

---

* *Malnutrition* refers to the physical effects of a diet intake that is inadequate in quantity and quality. *Undernutrition* refers to low food intake.

TABLE 8

Incidence of Protein-Energy Malnutrition (PEM) in Developing Countries[a,b]

Country	Year of survey	Age group covered (years)	Percentage suffering PEM		
			Severe	Moderate	Total
MSA[c]					
Guyana (national sample)	1971	0–4	1.3	30.8	32.1
Haiti	1971	0–5	6.0	25.0	31.0
El Salvador (national)	1967	0–4	3.1	22.9	26.0
Guatemala (national)	1967	0–4	5.9	26.5	32.4
Burundi	1972	0–5	2.2	28.7	30.4
Cameroon (Douala)	1973	0–5	4.4	36.4	40.8
Central African Empire	1972	0–5	3.0	36.4	39.4
Kenya	1968	—	1.0	25.0	26.0
Rwanda	1971	0–5	9.8	44.9	54.7
India		1–5	2.0	52.7	54.7
Non-MSA[c]					
Venezuela (national)	1971	1–6	0.9	14.5	15.4
Barbados (national)	1969	0–4	1.2	15.3	16.5
Jamaica (national)	1970	0–3	1.4	18.0	18.4
Brazil	1970	0–4	6.3	18.9	25.2
Colombia	1968	—	1.7	19.3	21.0

[a]From Bengoa and Donoso (1974); see also FAO (1977).
[b]The sample design of the different surveys included varied in coverage. Furthermore, the criteria of severity of malnutrition may also have varied as they could vary clinically or anthropometrically, or both.
[c]MSA, malnutrition – severe area.

Need for a Long-Range Coordinated Program. In order for the world's population to reach the levels predicted for 1985, food supplies (already inadequate) will have to be increased. In particular, supplies of animal products must be increased. Because the population is growing faster in the developing areas, to improve nutritional levels, total food supplies there will have to be increased more.

The U.S. Presidential Commission on World Hunger (1980) strongly recommended that this country take the lead in developing a long-range coordinated plan to meet the world food crisis. We "must do more to address the inequities that allow poverty and hunger to persist in the world." This would include short-term actions to alleviate immediate problems and longer-range measures to overcome poverty and inequity. Recommendations included family planning, increased crop yields (by increasing the cropland area and the yield per acre), improved protein supplies (new high-protein foods,

TABLE 9

Daily Calorie Availability per Capita by Income Groups, Brazil[a]

Income (cruzeiros per household per year)	Northeast[b]			
	Urban		Rural	
	% of households	Calories per capita/day	% of households	Calories per capita/day
Less than 100	9	1240	18	1500
100– 149	13	1500	14	1810
150– 249	26	2000	25	2140
250– 349	17	2320	13	1820
350– 499	14	2420	10	2228
500– 799	11	2860	11	2370
800–1199	5	3310	5	3380
1200–2499	4	4040	3	2870
More than 2500	1	4290	1	2900

[a]From the Getulio Vargas Foundation (1970); see also FAO (1977).
[b]Similar data are given for east and south Brazil in the original.

increased production of soybeans, peanuts, etc., increased animal production, fish-pond culture, increased fish catch, and development of fish protein concentrate), increased emphasis on nutritional needs, and research on protein foods from microorganisms, synthetic amino acids for enriching foods, and even recycling of waste paper as a source of carbohydrates. Reordering U.S. priorities would be difficult. Price- and trade-stabilizing agreements would be needed.

By the year 2000 the world's total food supply will have to be doubled or tripled in order to attain a reasonably adequate level of nutrition unless drastic measures to control the rate of growth of the population are taken. To achieve such production, new areas will have to be brought under cultivation and there will have to be an increased use of organic and inorganic fertilizers, better control of pests and disease, better seeds and improved methods of cultivation, greater use of irrigation, and double-cropping. In view of the increasing costs of energy, crop-production techniques that require a minimum of energy input will have to be developed. For animals, better and more scientific feeding, timely use of forage crops, control of animal diseases, and efficient breeding programs can increase productivity. Furthermore, a substantial increase in effective yields could be made by improved methods of processing and storage. A decrease in the birthrate would be the one most helpful step toward alleviating the problem. The alternative will probably be a famine of massive proportions. For other projections and recommendations, see USDA (1974, 1978) and International Food Policy Research Institute (1977).

Another point of view has been expressed by Tudge (1980), who opts for a future based on agricultural production not "geared primarily to short-term profits" but based on consumer societies that cook in a "rational" way. He believes less expenditure of energy would result if we produced more potatoes, cereals, and pulses (beans, lentils, chick-peas, and peas). Yet he allows room for food animals, on marginal lands or near human habitation.

The National Research Council (NRC) (1977) study of world food and nutrition problems recommended greatly expanded research in agricultural production but added that "research on the marketing, processing, preservation, and distribution of food could result in reduced food costs and could stimulate both production and consumption, particularly in the developing countries." The study's recommendations included research to reduce postharvest losses, market expansion, food reserves, crop information, and so on. There was hardly a word on new food sources or improving the nutritional value of processed foods! However, scientists are now deeply involved in using the new genetic technology to improve plant production (both quantitatively and qualitatively), to improve land and water use (with the aim of preventing salinity and erosion, providing better plant nutrition, and reducing costs), to improve photosynthetic efficiency, to increase nitrogen fixation (for better crop yields), and to reduce losses in quantity and quality due to insect, animal, and plant pests, bacteria, viruses, and fungi.

Norse (1979) believes that application of technology may not solve our food-production problems because "1) many subsistence farmers will not have access to the necessary production inputs, 2) the product may be too expensive for those most in need, and/or 3) the ecosystem cannot support the widespread use of intensive agricultural technology." He notes that the problem may be lack of demand, not of supply. There was adequate food in Ireland in the late 1840s to feed the population; famine occurred although higher-cost wheat was available, because low-cost potatoes had succumbed to a fungus disease.

Some developing nations are enslaved by a single cash crop that is often mainly used for export to secure foreign exchange. There is an urgent need to expand food production while continuing to produce the cash crops, in order to achieve a better-balanced agricultural economy.

Improving Protein Quality. The most serious problem in many developing countries is not the supply of calories but the supply of a sufficient amount and a balanced quality of protein. Several solutions to the problem of the lack of proteins are available. The development of high-yielding strains of corn with higher and better balanced amino acid content is one.

The *sea* did not appear to be a promising source of food to Bigwood (1967). The poor utilization of energy by plankton, etc. is the apparent reason. However, some believe that fish may represent one of the largest immediate potentials for new food production, particularly in certain regions. It has been reported that the fish catch from the oceans could be expanded by five to ten times. Japan and the Soviet Union have been leaders in developing the oceans as a source of food. Among the problems with fish are rapid spoilage and religious and social beliefs that limit fish consumption in some areas. Others believe that the inherent high capital costs of fish harvesting equipment, the uncertainty of political factors, high labor costs, and esthetic and traditional factors are forces that impede such developments. More important, food production in the sea is less efficient in terms of energy.

Synthetic foods offer one possibility of improving production (Pyke, 1970). Single cell proteins (SCP) have been produced from petroleum hydrocarbons, both crude and refined. Commercial development is still under way. How to incorporate such products into food so that it is acceptable to animals and people is a problem. Production of food yeast is a form of microbial protein production. Bigwood (1967) envisages proteins from microbial growth, plankton, and microbial growth on petroleum, and production of synthetic amino acids as supplements to proteins produced by conventional agriculture. Alcohol and fats can be produced from petroleum. Synthetic ascorbic acid and other vitamins are, of course, now common.

Pyke makes the useful proposal of creating synthetic foods that look and taste pleasant but that do not have any caloric value (for people on reducing diets). Increasing energy costs may slow these developments. Gershoff (1980) warns that synthetic and fortified foods may be less acceptable to consumers (because of their colors and odors) and may be less nutritious. Milner *et al.* (1980) also notes that production of synthetic protein still requires much research. Finally, *food enrichment* programs are needed. Beriberi and pellagra could be eliminated by proper enrichment of foods with the necessary vitamins. Enrichment with thiamin, riboflavin, niacin, calcium, and iron could be accomplished fairly simply. More difficult would be raising the protein content of low-protein foods. However, adding a missing amino acid is comparatively simple. Use of fish flour, egg powder, dried food yeast, skim milk, soybean or cottonseed flour, or other supplements has been suggested. Synthetic amino acids are also available. Synthetic methionine is widely fed to livestock and poultry in this country. Addition of lysine could raise the proportion of the wheat's usable protein from one half to two thirds. Addition of threonine would make grain protein almost as good in quality as the proteins of milk or meat. (See also Chapter 6.)

Incaparina is an example of a new food of high nutritional value intended for areas of poor diet, primarily for babies. It is a mixture of corn, sorghum, cottonseed flour, dried yeast, and synthetic vitamin A and contains about 26% protein. It is being distributed in Central America, particularly in Guatemala and Colombia. CSM is a U.S.-produced relief food made of 70% gelatinized corn flour, 25% soy flour, and 5% nonfat dry milk powder. It has about 25% protein. Over a billion pounds have been shipped to relief areas. Bal-Amul is a mixture of soy and milk protein that is being marketed in India as a baby food. Solein, also used as a milk substitute, is a similar Brazilian product. Protein-enriched soft drinks have found a market in several areas. Vitasoy is a soybean-based beverage that has captured 25% of the soft drink market of Hong Kong. Saci is a similar and successful product in Brazil, as is Puma in Guyana. Similar products are being developed in other countries.

Unfortunately, problems arise because the powdered food substitutes have to be mixed with water before use and uncontaminated water is not easily available in many areas. The result is that the child receives contaminated food and suffers from gastrointestinal diseases of varying degrees of seriousness, even leading to death. Jelliffe and Jelliffe (1980) recommend breastfeeding in developing countries to reduce infantile obesity, cows' milk allergy, early marasmus, and diarrheal disease.

The real need for the food industry in developing countries is to produce commercially low-cost, high-protein, high-calorie, high-prestige foods, preferably from local raw materials. They should suit local needs and ecological circumstances and must avoid interference with the local lactation pattern.

U.S. Food Production and Consumption

Production

Excellent data on food production and consumption are available for the United States (Burk, 1961; U.S. Department of Agriculture, 1973, 1979, 1981a). Consumption may be calculated on the basis of nutritive value (calories, for example) or of pounds, or on a food-consumption or food-use index, both of which are based on price and volume. The trends since 1910 are given in Fig. 15. Note the steady increases based on price and the decrease based on calories or pounds. The decrease is due partially to a shift from rural and outdoor occupations to urban life with its lesser requirement for calories.

Consumption

In 1979, a U.S. family of four consumed about 3 tons of food per year. Meat, poultry, and fish amounted to nearly half a ton and dairy products

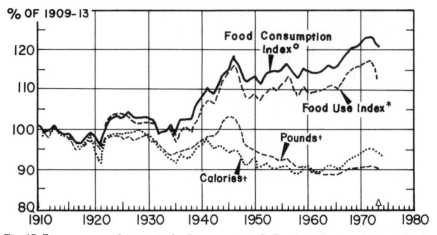

Fig. 15. Four measures of per capita food consumption. O, Retail-weight equivalent, weighted by constant retail prices; *, Farm-weight equivalent, weighted by prices received by farmers (or equivalent); index adjusted to level of food-consumption index in 1924 (1924 = 103.8); ‡, Available for consumption at retail level; †, Retail-weight equivalent. Δ, Preliminary. (From USDA, 1981.)

close to two thirds of a ton. Fruits and vegetables accounted for well over half a ton, and flour and cereal products, sugar, potatoes, fats, and oils over four fifths of a ton. This gave a daily per capita calorie intake of about 3200, which is nearly 20% above the recommended level (see USDA, 1966, 1973, 1979).

Changes in per capita consumption by specific commodities since 1975 are shown in Table 10 (on a retail-weight equivalent). Consumption of dairy products, potatoes and sweet potatoes, and coffee, tea, and cocoa has declined, while that of meats, fish, and poultry, fats, processed fruits, frozen and canned vegetables, and sugars and syrups has increased. Results since 1965 are shown on a price-weighted basis in Table 11. Canned and frozen vegetables, beans, peas, and nuts show relatively large increases, as do potatoes and sweet potatoes. This last result is not due to a sudden increase in potato consumption, but to a shift from inexpensive bulk potatoes to more expensive frozen, processed, or canned potatoes (which are, of course, more easily prepared).

Diet. The U.S. diet has changed very little in the last 13 years according to Marston and Welsh (1981). The nutrients available per capita per day in 1967 and 1980 were as follows: calories, 3240 vs. 3520; protein, 99 vs. 109 gm; fat, 152 vs. 168 gm; carbohydrates, 374 vs. 406 gm; calcium, 947 vs. 891 mg; phosphorus, 1529 vs. 1528 mg; iron, 16.4 vs. 17.6 mg; magnesium, 341 vs. 343 mg; vitamin A, 7900 vs. 8400 IU; thiamin, 1.9 vs. 2.2 mg; riboflavin, 2.3 vs. 2.4 mg; niacin, 23.2 vs. 26.8 mg; vitamin B_6, 1.9 vs. 2.0 mg; vitamin

TABLE 10
Approximate Consumption of Food per Capita in the United States[a,b]

Item	1975	1979
Meats, fish, and poultry	224	236
Dairy products, including butter	350	345
Eggs	35	36
Fats and oils, excluding butter	52	57
Fruit		
Fresh	81	81
Processed	57	58
Vegetables		
Fresh	141	144
Canned	52	53
Frozen	10	11
Potatoes and sweet potatoes	83	82
Dry beans and peas, nuts, soya products	18	18
Flour and cereal products	139	150
Sugars and syrups	121	137
Coffee, tea, cocoa	13	12
Retail-weight equivalent	1415	1463

[a]Data from USDA (1981a, p. 554).
[b]Retail-weight equivalent in pounds. 1967–1969 = 100.

TABLE 11
Index of per Capita Food Consumption in the United States[a,b]

Item	1965	1970	1975	1979
Meat, fish, and poultry (including fat pork cuts)	99	105	102	106
Eggs	98	97	87	88
Dairy products, including butter	103	99	99	101
Fats and oils, excluding fat pork cuts and butter	96	108	110	120
Fruits				
Fresh	101	101	105	106
Processed	98	103	112	110
Vegetables				
Fresh	100	101	101	103
Canned	95	104	109	109
Frozen	89	101	106	127
Potatoes and sweet potatoes	99	108	109	115
Beans, peas, nuts	100	98	107	111
Flour and cereal products	101	98	96	105
Sugar and syrups	98	106	104	117
Coffee, tea, and cocoa	99	93	89	84
All food	98	102	106	106

[a]Data from USDA (1981a, p. 554).
[b]Price-weighted basis; 1967–1969 = 100.

B_{12}, 9.4 vs. 9.5 μg; ascorbic acid, 105 vs. 123 mg. They also note that zinc intake is 12.5 mg per capita per day (about the same as in 1900).

Prices and Value

The increases in retail and farm food prices continue. As of 1979 they were 251% and 227% of 1967, respectively (USDA, 1981a). Retail and wholesale prices have been less influenced by economic recession than farm prices. Food prices lagged behind the consumer price index in the 1920s and 1930s but since then generally have followed the same pattern.

The farm value of U.S. foods amounted to $76.8 billion in 1979 (USDA, 1980a). The marketing bill was $162 billion and civilian expenditures for foods amounted to $238.8 billion. This compares with $27.1, $54.0 and $81.1 billion, respectively, in 1965! By the last quarter of 1980 personal consumption figures were nearly $300 billion (Gallo, 1981). The chief components of the increases in food prices were due to retail food price changes and increased marketing services. Some of the added service cost is due to technological changes. Wheat no longer dries in the stack, so the elevator must dry and condition it. Potatoes dug by machine do not harden in the field and must be more carefully handled in storage. The percentage of costs devoted to marketing services is still increasing.

U.S. expenditures for food represent about 14% of *total* consumer expenditures. In contrast, Canada spends about 16%, France over 20%, Germany over 22%, South Africa 25%, Korea over 45%, and the Philippines 59%. The percentage spent for food in the United States has been decreasing for many years. Gallo (1981) notes that the percentage of *disposable* income spent for food dropped to 16.6% in 1980 (12.2% at home and 4.4% away from home). This compares with 24% in 1929 (20% at home and 4% away from home).

Total food consumption, of course, follows population. Figure 16 shows a slight rise in the retail weight consumption using constant retail prices as index weights of 1967 data.

Production of Nutrients

Available food or actual food consumption in the United States is given in Figs. 16 and 17. The data (Fig. 18) indicate the improved supply of protein, fat, carbohydrate, thiamin, niacin, and ascorbic acid in the U.S. diet. Ascorbic acid and vitamin A are about the same as in the 1909–1913 period. Thiamin intake has increased since 1940, when bread enrichment with thiamin was instituted. Per capita consumption of crop products shows a marked rise (Fig. 17). Consumption of animal products has fluctuated but overall has changed

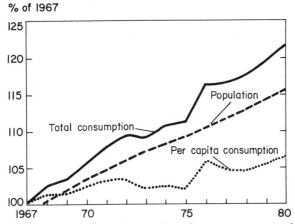

Fig. 16. Total food consumption based on retail weight using constant retail prices as index weights. Civilian population on July 1, for 50 states. 1980 projected. (From USDA, 1981b.)

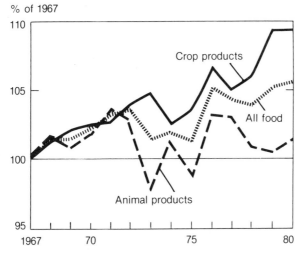

1980 preliminary.

Fig. 17. Per capita consumption of crop and animal products. Civilian consumption (using constant retail prices as index weights). (From USDA, 1981b.)

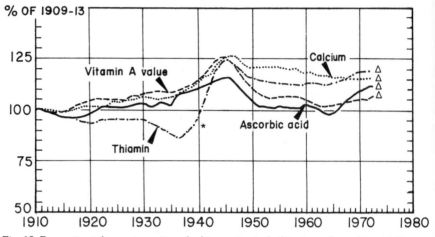

Fig. 18. Per capita civilian consumption of calcium, vitamin A, thiamin, and ascorbic acid. Five-year moving average. ★, Enrichment initiated; Δ, preliminary. (From USDA, 1981a.)

little. During the Depression of the 1930s and in 1973–1975, consumption of animal products declined more than that of crop products. In Figs. 19 and 20 other changes in the U.S. diet are obvious. Figure 19 shows the long-term decreasing consumption of carbohydrates and increasing consumption of fat. Figure 20 indicates that less of the protein in the U.S. diet is derived from crop sources than from animal sources.

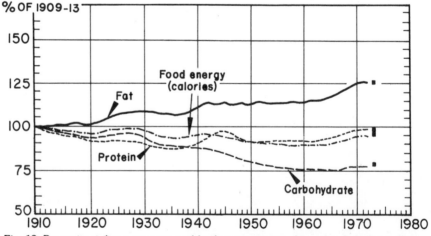

Fig. 19. Per capita civilian consumption of food energy, protein, fat, and carbohydrates. Five-year moving average. ■, Preliminary. (From USDA, 1981a.)

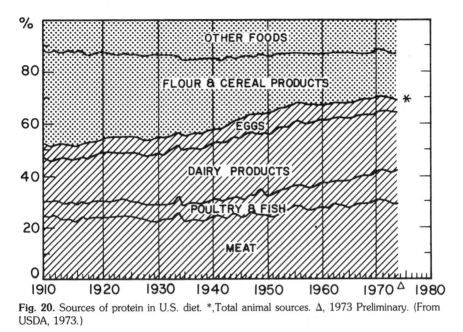

Fig. 20. Sources of protein in U.S. diet. *,Total animal sources. Δ, 1973 Preliminary. (From USDA, 1973.)

Production by Commodities

Similar data are given in Fig. 21 for several commodities. Note especially the changes in sugar, coffee, poultry, and vegetable oil consumption. Poultry prices have decreased as per capita consumption increased. Broiler consumption has been the chief reason for the increase. Per capita consumption of *fresh* fruits and vegetables has increased, but not as much as for *processed* fruits and vegetables. Apparently the consumer is willing to pay more for the added convenience of processed fruits and vegetables. Even restaurants and institutions are using increasing amounts of prepared food to reduce labor costs. Increasing amounts of both hot and cold food are being dispensed in vending machines.

Changes in consumption continue (Smallwood et al. 1981). Between 1965 and 1977, decreases in expenditures were as follows: cereal and bakery products (4%), dairy products (7%), meats, poultry, and eggs (14%), fruits (18%), vegetables (4%), and fats and oils (1%). Sugars and sweets rose 31% and other foods 29% (cheeses, fish, and processed vegetables).

Prices of cereal and bakery products have risen steadily since 1909, while per capita consumption has declined. These trends may not be correlated. Carbohydrate consumption appears to be decreasing because of greater attention to weight control by Americans. Per capita consumption of liquid sweeteners, however, has increased since 1967. The actual price has risen but the deflated retail price has remained remarkably constant.

Per Capita Consumption of Selected
Animal Products

% of 1967

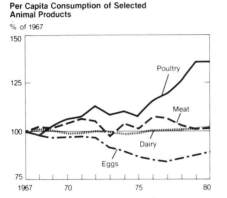

Per Capita Consumption of Selected Crop Products

% of 1967

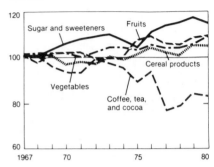

Per Capita Consumption of Fats and Oils

Per Capita Consumption of Fruits and Vegetables

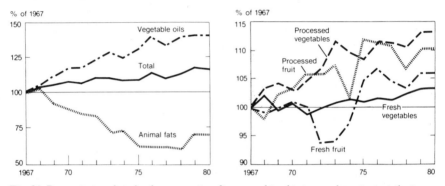

Fig. 21. Per capita trends in food consumption. Items combined in terms of constant retail prices. Butter included with both dairy products and fats and oils. Vegetables excludes potatoes, peas, and beans. Fruits includes melons. 1980 Preliminary. (From USDA, 1981.)

In summary, long-term changes in the U.S. diet are occurring. Some of these changes may be related to increasing prices but others are surely due to the demand for convenience foods, to greater care in calorie intake, and to increased personal income.

Income and Food

Personal disposable income has increased remarkably since 1960. At the same time, the percentage of income spent on food has decreased! For families of high income (e.g., over $20,000), as little as 10% may be spent on food. Families with children and a low income (e.g., $5,000 or less) spend 30% or more on food (USDA, 1980b). Rural families spend slightly more of their income on food than do urban families. Families in the southern and northeastern United States spend a slightly greater percentage of their income on food than do those in the western or north central regions. During World War II—when a high percentage of the population was employed, often at higher wage levels than they were accustomed to—a greater percentage of disposable income was spent on food. Other consumer goods, particularly automobiles, were unavailable or in short supply. Many housewives worked and had money to spend and less time to prepare food. They therefore bought precooked or convenience foods or ate away from home a greater percentage of the time. All of these factors increased the percentage of income spent for food. It is also interesting to note that Engel's law was broken during the war years. (Engel's law states that as the income increases a smaller percentage is spent on food.) In general this is true, but when people are faced with surplus income, as occurred during World War II, the law may not hold true. Since 1960 (to 1979), food consumption increased by only 33 pounds, or 2.3%. However, the total food-consumption index (weighted to 1967–1969 prices) rose 8.5%. This means that Americans have increased the proportion of higher-valued foods in their diet (USDA, 1981a).

The U.S. consumer works less for more and better food than do consumers in any other country. An industrial worker in this country in the mid-1960s could purchase a meal for four persons with 1 hour's pay. In England it required 2 hours, in Austria 4, in France 4.5, and in Italy 5. In 1965 we spent about 18% of our take-home pay on food, compared to 26% in 1947–1949. Consumers in Great Britain in 1965 spent nearly 30%, Russians more than 40%. In the developing countries, food costs take half or more of the income (USDA, 1966). The situation has improved somewhat for the developed countries in recent years.

As income increases, total food expenditures increase. One reason is that the amount spent for food away from home increases, and the food received without cost or produced at home decreases. Thus the income group markedly influences the distribution of the consumer's dollar. Higher-income groups spend more on soup, vegetables, potatoes, sweet potatoes, fruit, bakery products, meat, cream, and cheese than does the lower-income group. They buy less flour, fats and oils, eggs, and sugar and sweets.

About 90% of the food consumed in this country is produced in the United States. Some foods—coffee, tea, and cocoa, for example—are not grown in the continental United States.

Long-Range Trends

Bennett and Peirce (1961) estimated that in the 1970s U.S. consumption of calories would decline slightly; it did not. Their predictions of some other diet changes likewise did not occur.

They attribute the long-range reduction in the number of calories being consumed to five factors: less arduous physical labor, fewer hours worked per week, less exposure to cold weather, conscious control of body weight, and minimization of wastes in wholesaling and retailing. The data of the U.S. Department of Commerce (1980) show increases in 1979 consumption (compared to 1960) of red meat (111%), ready-to-cook chicken (114%), fats and oils (112%), frozen fruit and juices (101%), and canned and frozen vegetables (118% and 137%, respectively). Declines may be noted for total milk fat solids (84%), refined sugar (94%), and dry edible beans (81%).

Convenience. The present American food market may be defined as a market for convenience. Instead of buying a chicken, the American household is now demanding frozen chicken livers separately from canned breast chicken or from dehydrated chicken soup. Naturally they have to pay more for the convenience that has been built into this type of food. The demand for convenience may be due to the fact that more and more housewives now have jobs (about 50% of them were working in 1980 compared with only 24% in 1941) and also to the large increase in the birthrate in the immediate post–World War II years, which means that the American housewife had less time for work in the kitchen.

This increase in the number of convenience items is easily understood. There is less home baking of bread and home canning, which were formerly very common. As much as 20% of the fresh meat sold is trimmed, prepackaged, and prelabeled. An increasing percentage of coffee is sold as instant coffee, and completely prepared TV dinners are now common and are available in a variety of different styles.

Another illustration of this is the spectacular growth of the frozen-food industry, starting in 1947. Vegetables, fruits, poultry and seafood are now available in frozen or canned packages. Production of concentrated and frozen orange juice has increased, and at the same time fresh orange consumption has been dropping. Citrus juice, mostly orange, now constitutes 65% of the total produced. Of the noncitrus juices, tomato and pineapple are the most important.

Items that formerly were not looked on with favor, such as frozen strawberries, frozen red meat, and frozen broccoli, have become accepted as convenience items. In addition, the canning industry is still growing. Canned food consumption generally increases as the income increases up to about $5000, but above this many canned items do not increase. For example, the consumption of canned pears increases steadily with income, whereas canned fruit cocktail consumption decreases sharply in the upper income brackets. The canned food industry has held its own on a per capita basis but has not expanded any faster than the population, whereas the frozen food industry has. The one exception to this is in the baby food industry. Canned strained baby foods have become a "must" for American families, and sales have increased enormously since 1935.

Research and Development. One result of emphasis on research and development is the large number of new food products: cake mixes, low-calorie salad dressings, numerous diet (low sodium or low calorie) foods, breakfast foods, combination frozen foods, frozen buttered vegetables, and so on. Of the research-and-development dollar of food processors, 10% or more is going to new product development. Two thirds of the food products that will be consumed in 1984 had not yet been developed in 1965.

In introducing new convenience foods, Harp and Miller (1965) found that cost per serving, degree of competition from similar products, importance of the food group to the consumer (sales volume for all forms of the item), availability as a convenience food, success of similar items, and degree of convenience offered are all important in predicting sales success. They did not measure the effect of food quality or of promotional activities.

The increase in convenience items has also been accompanied by an increase in the cost of service. The labor, storage, transportation, refrigeration, and assembly lines all have taken a large share of the food dollar. The food industry appears to be expanding its gross business largely by selling more and more "extras." Inflation, of course, has increased prices, but the consumer is paying for the increased services.

Supermarkets. The greatest increase has been in the retail store, where the amount taken by the store has risen spectacularly. This is, of course, because the American store has become much different than it was some years ago. The American supermarket needs parking space and this costs money. In addition, the amount of home deliveries has decreased in recent years. This means that the consumer purchases larger quantities in the retail store that must be taken along home. Thus the wire carriages have tended to increase in size. Electric-eye doors have become common so that the packaged food can be wheeled out easily. In addition, because the consumer spends more time in the store, it has not only had to be air conditioned but rest rooms, fluorescent lighting, lunch counters, and a variety of other services have been provided. The American supermarket may supply 5000 or more different items (compared to less than 1000 before 1930).

The question whether the supermarket has led to a great deal (too much?) of impulse buying should be considered. The self-service of the supermarket undoubtedly is a wonderful opportunity to pick and choose. Surveys have shown that consumers do buy items which they had not planned to buy when they entered the store. More important, the high-margin-of-profit items tended to be bought with the least planning. As much as two thirds of the purchases of candy, dessert mixes and chewing gums, pie and pretzels seem to be completely unplanned! It can be argued that the unplanned purchases are taken out of future spending, since the total amount of budget that can be spent on food is limited by the family income and by the appetities of the family. However, it may be that consumers spend more for food because they eliminate nonfood purchases that they are not so tempted to buy because of the less frequent exposure to them as compared to the exposure to impulse buying in a supermarket.

The ethics of supermarket sales practices have often been questioned and have been the subject of a congressional hearing. The charge that the food industry labels and advertises its merchandise in a deceiving manner was countered by a food industry spokesman who asserted that the legally necessary weight or volume information is given on the label and that the consumer is not "rational." The term "jumbo half-quart" was said to fill a psychological need of the consumer for plentitude! That the congressional hearings did some good is indicated by the fact that one major chain inspected its large "economy-size" packages and found some that did not cost less per unit weight. There has been a marked improvement in labeling practices. Most retail food stores now give complete information on the age of certain perishables (e.g., milk) rather than furnishing the information, as formerly, in codes that the consumer has difficulty deciphering.

Basis of Consumer Purchases

How to explain the basis of consumer purchases has been studied by Salathe and Buse (1979) using the 1965 U.S. food survey data (USDA, 1967). Household size and composition are important factors. Newborn babies obviously have less impact than adults. Male and female children differ in their scale of food values from male and female adults. Age is also important. Elderly females have less impact on household expenditures for total food, vegetables, and beef and pork than do females ages 20–55 years. The increasing age of the U.S. population has had a negative impact on per capita consumption of total food, beef and pork, vegetables, grain products, and dairy products. Food expenditures per person tend to decline as more adults are added to the family. One reason for this is the opportunity to benefit from large purchases as family size increases.

Socioeconomic and demographic factors are also important: location of household (higher expenditures in the Northeast, lowest in the South); urban vs. rural (more spent on total food, grain products, and beef and pork, less on vegetables, dairy products, and fruits in urban areas), education and race (as education increases, the marginal propensity to spend for total food, vegetables, beef and pork declines; black households have a greater tendency to spend for food than do white, etc.).

U.S. Food Problems

Although famine is unknown in this country, many people survive on inadequate diets. As late as 1940 there were 2123 deaths from pellagra, 63 from beriberi, 26 from scurvy, and 161 from rickets. For every death, of course, a great many more persons are affected. In addition, some deaths are reported as being due to other diseases rather than to malnutrition. This was clearly established during World War II in surveys by the National Research Council (1943). Low income families still sometimes have an average daily caloric intake below the recommended allowances or suffer from dietary deficiencies, due to lack of income. There are estimates that 25 million Americans are currently underfed. Lack of knowledge of simple dietary rules and lack of understanding of the importance of good nutrition are also involved.

In a nationwide survey made in 1965 the USDA (1967, 1969a) found that average diets in 90–100% of the sex–age groups were above the recommended levels for calories, protein, vitamin A value, thiamin, riboflavin, and ascorbic acid. However, calcium and iron intake were more often below the recommended levels. About 30% of the diets of girls and women were deficient in those two elements. About 50% of infants and children under 3 years of age had calcium- and iron-deficient diets. In general the diets of U.S. males provide more nutrients than do those of females. For the under-$3000 annual income group and for persons in the southern states, ascorbic acid and vitamin A as well as calcium and iron intake were most often deficient. The 1965 survey has been updated but complete data have not yet been published. The greater susceptibility and lower resistance to infections of people in a deficiency status should be emphasized. A Canadian survey (Adamson et al., 1945) stated, "The poor nutritional status of the people of Newfoundland may well be in large part responsible for their impaired health and efficiency."

United Kingdom and European
Food Consumption

Changes in food consumption similar to those of the United States have taken place in the United Kingdom, according to Drummond and Wilbraham (1958) and Burnett (1966). The percentage of calories derived from grain and potatoes has declined steadily since 1880. However, it is the type of carbohydrate consumed that has changed. Great Britain has the highest per capita consumption of sugar of any country. Though bread consumption has decreased, the decrease is mainly in white unwrapped bread. Consumption of brown and whole-meal bread has been maintained. With the decreasing overall bread consumption, there has also been a decrease in butter and margarine use (Great Britain, 1978).

Similar data could be cited for most countries of western Europe. For example, Glatzel (1961) has analyzed the changes in food habits in Germany since 1880. In Germany, bread consumption has decreased from 300 kg per capita per year to 200 kg in 1900 and 70 kg in 1961. Fruit consumption has increased 40% since World War II. Reduced costs of production have made luxury items accessible to all groups.

In Great Britain there are also geographical differences in nutrient intake. The Scottish diet provides less energy and vitamins than that of other regions but the differences are not large. Scots also eat more bread and meat but less vegetables and fruit. The diet in Scotland and Wales provides less vitamin C than in other areas. Income does not have a pronounced effect on the nutritional adequacy of British diets. Allen (1968) shows the varied regional tastes of Great Britain: more stewed beef and spirits (Scotland), more pickles and peas (Midlands), less baked beans or blue cheese but more canned peas (Northwest).

Convenience foods have also increased their share of the food market in Great Britain. In 1970 only 3% of households had a deep-freezer. In 1975 23% had one. In 1956 only 8% of households had refrigerators; in 1975 88% did. Because convenience foods are relatively expensive sources of nutrients, they accounted for less of the energy and the protein in the household diet. Families with children spent proportionately more on convenience foods than those without children.

As in the United States, consumption of poultry has increased steadily (sixfold between 1956 and 1967). Poultry has moved from a luxury to a conventional necessity in the British diet. This is, of course, due to the mass production of poultry, which has reduced the cost per pound.

REFERENCES

Adamson, J. D., Jolliffe, N., et al. (1945). Medical survey of nutrition of Newfoundland. Can. Med. Assoc. J. 52, 227–250.

Allen, D. E. (1968). "British Tastes. An Enquiry into the Likes and Dislikes of the Regional Consumer." Hutchinson, London.

Anonymous (1968). Strategy Conquest Hunger, Proc. Symp. Rockefeller Found., New York.

Anonymous (1981). In "Fertility Decline in the Less Developed Countries," AAAS Symposium. Praeger, New York.

Bengoa, J. M., and Donoso, G. (1974). Prevalence of protein-calorie malnutrition, 1963 to 1973. *Protein–Calorie Advis. Group, U.N.* **4**(1), 24–35.

Bennett, M. K. (1963). Longer and shorter views of the Malthusian prospect. *Food Res. Inst. Stud.* **4**, 3–11.

Bennett, M. K., and Peirce, R. H. (1961). Change in American national diet, 1879–1959. *Food Res. Inst. Stud.* **2**, 95–119.

Bigwood, E. J. (1967). De l'avenir de notre alimentation. Aperçu sur l'an 2000. *Ind. Alliment. Agric.* **84**, 845–857.

Biswas, M. R., and Biswas, A. K. (1979). "Food, Climate, and Man." Wiley, New York.

Borgstrom, G. (1965). "The Hungry Planet. The Modern World at the Edge of Famine." Macmillan, New York.

Borgstrom, G. (1980). The food and people dilemma. *In* "Nutrition, Food and Man. An Interdisciplinary Perspective" (P. B. Pearson and J. R. Greenwell, eds.), pp. 71–81. Univ. of Arizona Press, Tucson.

Brooks, R. R. R. (1970). People versus food. *Sat. Rev.* Sept. 5, p. 10–14, 33.

Brown, L. R. (1981). World population growth, soil erosion, and food security. *Science* **214**, 995–1002.

Burk, M. C. (1961). Trends and patterns in U.S. food consumption. *U.S., Dep. Agric., Agric. Handb.* No. 214, 1–123.

Burnett, J. (1966). "Plenty and Want; A Social History of Diet in England from 1815 to the Present Day." Nelson, London.

Campbell, K. (1979). "Food for the Future." Univ. of Nebraska Press, Lincoln.

Deevey, E. S., Jr. (1960). The human population. *Sci. Am.* **203**, 194–204.

Drummond, J. C., and Wilbraham, A. (1958). "The Englishman's Food; A History of Five Centuries of English Diet." Cape, London.

Ehrlich, P. E., and Ehrlich, A. H. (1970). "Population, Resources, Environment. Issues in Human Ecology." Freeman, San Francisco, California.

FAO (1977). "The Fourth World Food Survey," Statistical Series 11. Food Agric. Organ. U.N., Rome.

FAO (1979a). "FAO Commodity Review and Outlook 1979–1980." Food Agric. Organ. U.N., Rome.

FAO (1979b). "Agriculture: toward 2000." Food Agric. Organ. U.N., Rome.

FAO (1980a). "The State of Food and Agriculture 1979." Food Agric. Organ. U.N., Rome.

FAO (1980b). "FAO Production Yearbook 1979," Vol. 33. Food Agric. Organ. U.N., Rome.

Gallo, A. E. (1981). Food spending and income. *Natl. Food Rev.* No. 14, pp. 2–4.

Gershoff, S. N. (1980). The fortification of foods. *In* "Nutrition, Food and Man. An Interdisciplinary Perspective" (P. B. Pearson and J. R. Greenwell, eds.), pp. 65–70. Univ. of Arizona Press, Tucson.

Glatzel, H. (1961). Entwicklungstendenzen in der Kostwahl und ihre Auswirkungen auf die Ernährungssituation. *Ernaehr.-Umsch.* **8**, 231–233.

Great Britain. Ministry of Agriculture, Fisheries and Food (1978). "Household Food Consumption and Expenditure: 1977." HM Stationery Off., London. (See also earlier reports.)

Hardin, G. (1974). Lifeboat ethics: the case against helping the poor. *Psychol. Today* **8**(4), 36–43.

Harp, H. H., and Miller, M. B. (1965). Convenience foods. The relationship between sales volume and factors influencing demand. *U.S., Econ. Res. Ser., Agric. Econ. Rep.* **81**, 1–21.

Harrison, G. G. (1980). Strategies for solving world food problems. *In* "Nutrition, Food, and Man. An Interdisciplinary Perspective" (P. B. Pearson and J. R. Greenwell, eds.), pp. 1–10. Univ. of Arizona Press, Tucson.

Hayes, H. T. P. (1981). "Three Levels of Time." Dutton, New York.

International Food Policy Research Institute (1977). Food needs of the developing countries— projections of production and consumption to 1990. IFPRI *Res. Rep.* No. 3, pp. 1–157.

Jelliffe, D. R., and Jelliffe, E. F. P. (1980). Breast feeding and infant nutrition. *In* "Nutrition, Food and Man. An Interdisciplinary Perspective" (P. B. Pearson and J. R. Greenwell, eds.), pp. 11–20. Univ. of Arizona Press, Tucson.

Manocha, S. L. (1975). "Nutrition and Our Overpopulated Planet." Thomas, Springfield, Illinois.

Marston, R. M., and Welsh, S. O. (1981). Nutrient content of the national food supply. *Natl. Food Rev.* No. 13, pp. 19–21.

Milner, M., Scrimshaw, N. S., and Wang, D. I. C. (1980). World food requirements and the search for new protein sources. *In* "Nutrition, Food and Man. An Interdisciplinary Perspective" (P. B. Pearson and J. R. Greenwell, eds.), pp. 88–104. Univ. of Arizona Press, Tucson.

Murdock, W. M., and Oaten, A. (1975). Population and food: metaphors and the reality. *BioScience* **25,** 561–567.

Myrdal, G. (1970). "The Challenge of World Poverty: A World Anti-Poverty Program in Outline." Pantheon Books, New York.

National Research Council. Commission on International Relations (1977). "World Food and Nutrition Study: the Potential Contributions of Research," Publ. No. 2628. Nat. Acad. Sci., Washington, D.C.

Norse, D. (1979). Natural resources, development strategies, and the world food problem. *In* "Food, Climate, and Man" (M. R. Biswas and A. K. Biswas, eds.), pp. 12–15. Wiley, New York.

Paddock, W., and Paddock, P. (1967). "Famine—1975! America's Decision: Who Will Survive? Little, Brown, Boston, Massachusetts.

Poleman, T. T. (1975). World food: a perspective. *In* "Food: Politics, Economics, Nutrition and Research" (P. H. Abelson, ed.), pp. 8–16. Am. Assoc. Adv. Sci., Washington, D.C.

Pyke, M. (1970). "Synthetic Food." Murray, London.

Salathe, L. E., and Buse, R. C. (1979). Household food consumption patterns in the United States. *U.S. Dep. Agric., Tech. Bull.* No. 1587, pp. 1–27.

Smallwood, D., Blaylock, J., and Zellner, J. (1981). Factors influencing food choice. *Natl. Food Rev.* No. 14, pp. 20–22.

Tudge, C. (1980). "Future Food: Politics, Philosophy, and Recipes for the 21st Century." Harmony Books, New York.

United Nations (1978). "Statistical Yearbook for Asia and the Pacific." Econ. Soc. Comm. Asia Pac., Bangkok.

U.S. Department of Agriculture (1966). "Protecting our Food." U.S. Gov. Print. Off., Washington, D.C.

U.S. Department of Agriculture (1967). Food consumption of households in the United States, Spring 1965. *U.S. Household Food Consumption Surv., 1965–1966 Rep.* No. 1, pp. 1–211.

U.S. Department of Agriculture (1973). Food. Consumption, prices, expenditures. Supplement for 1973. *U.S., Econ. Res. Ser., Agric. Econ. Rep.* No. 138, Suppl., pp. 1–14.

U.S. Department of Agriculture (1974). The world food situation and prospects for 1985. *U.S., Dep. Agric., Foreign Agric. Econ. Rep.* No. 98, pp. 1–90.

U.S. Department of Agriculture (1977). "Asia and Oceania Agricultural Situation. Review of 1976 and Outlook for 1977." Washington, D.C.

U.S. Department of Agriculture (1978). Alternative future for world food in 1985. *U.S., Dep. Agric., Foreign Agric. Econ. Rep.* No. 151, pp. 1–96.

U.S. Department of Agriculture (1979). Food. Consumption, prices, expenditures. Supplement for 1979. *U.S., Econ. Res. Ser. Agric. Econ. Rep.* No. 138, Suppl., pp. 1–98.

U.S. Department of Agriculture (1980a). "Agricultural Statistics 1980." U.S. Gov. Print. Off., Washington, D.C.

U.S. Department of Agriculture (1980b). Global food assessment, 1980. *U.S., Dep. Agric., Foreign Agric. Econ. Rep.* No. 159, pp. 1–119.

U.S. Department of Agriculture (1981a). Food. Consumption, prices, and expenditures. U.S., *Dep. Agric., Econ. Stat. Div., Stat. Bull.* No. 656, pp. 1–90.

U.S. Department of Agriculture (1981b). Handbook of agricultural charts 1981. *U.S., Dep. Agric., Agric. Handb.* No. 592, pp. 1–99.

U.S. Department of Commerce (1980). "Statistical Abstract of the United States, 1980," 101st ed. U.S. Gov. Print. Off., Washington, D.C.

U.S. Presidential Commission on World Hunger (1980). "Overcoming World Hunger: The Challenge Ahead." U.S. Gov. Print. Off., Washington, D.C.

Chapter 3

Food Habits, Taboos, and Quality Attributes

Food science and technology is concerned with the transformation of natural raw materials (e.g., fruits and vegetables, milk, eggs, and meats) and other ingredients (condiments, colors, and additives) into safe, palatable, nutritious foods. Before discussing this main topic, we must consider the basic principles and the practical applications relating to the concerns and attitudes of people toward their food. (In this text, the word *food* will be understood to include most beverages.)

First, just what does the word *food* mean to us? What factors determine whether a given product will be consumed by few, some, or all members of the population? This question involves basic genetic, physiological, and psychological factors. Then, what are the more complex cultural, religious, geographical, and climatic influences that affect the acceptance or rejection of raw or processed products as foods to eat and enjoy? Further, what constitutes quality in food: what makes a product highly prized, only marginally acceptable, or barely tolerated?

To answer these questions, we will examine what current knowledge reveals and, as far as possible, what remains to be discovered. The objective is to provide an introduction for our main concern: the principles and practices of commercial food processing and preservation, which we will consider in later chapters.

Food Habits

Our food habits are a part of our cultural and emotional lives. They are a major factor in determining the quality of life and some have become rituals in our daily living.

Origins of Food Habits

Almost all animals eat a large range of foods. Whether herbivore, carnivore, or omnivore, all animals, including human beings, must learn which food is nutritious and which is poisonous. Thus we have sensory receptors to ensure recognition of salty and sweet foods; to avoid poisonous foods we rely on the bitter taste (Garcia and Hankins, 1975). Even babies avoid a bitter taste; weaning is promoted in some societies by smearing a bitter-tasting substance on the mother's breast. Learning about the positive effects of a food is slower than learning about negative effects, according to Seward and Greathouse (1973).

Food acceptance is based on human physiology, sensory perception, and personal attitudes (including cultural patterns; see Rozin, 1976). In the case of physiology, the acceptance of the food depends on hunger and appetite and varies with how long the food is visible. (This is discussed further in Chapter 5.) The influence of attitudes depends on whether the food experience has been repeated seldom or many times. Even the eating situation itself can be a factor, such as manner of serving, decor, social groupings, and extraneous noise. Genetic factors may also have profound effects on the food habits of individuals, but these have not yet been studied sufficiently to draw any conclusions. Humans have inherited a catholicity of food habits from vegetarian to carnivorous (Le Gros Clark, 1968). Their origins are a blend of speculation and inference (Todhunter, 1979). We have well-developed likes and dislikes for certain foods, in addition to those that can be attributed solely to purity and safety, convenience, functional properties, and nutritional value. De gustibus non est disputandum (There is no disputing about taste) is a fundamental statement of the individuality of taste preference.

A model for all the factors influencing food acceptance is given in Fig. 22.

Social Factors

The ways in which individuals or groups select, consume, and utilize their food constitute their food habits. Not everyone in a group has the same food habits. Families may form subgroups with various deviations from the group norm. These deviations include variations in food production, storage, processing, distribution, and consumption (Yudkin and McKenzie, 1964).

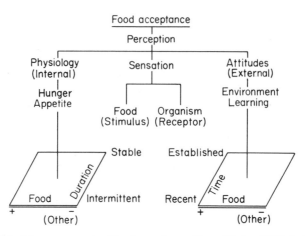

Fig. 22. Model of the components of food acceptance. (From Pilgrim, 1957a,b. © *Am. J. Clin. Nutr.*, American Society for Clinical Nutrition.)

Changes in food habits may be more difficult to effect than shifts in religious sects. Food habits of people who enjoy food are apparently more difficult to change than those of populations where food is difficult to procure or is simply one element of a rigid disciplinary system. They are especially resistant to change in societies where they have been built into the child's early experience in a particular way. However, when a society changes its food habits the younger members of the family tend to do so first.

Worldwide changes in food habits are possible: consider the introduction of corn, peppers, potatoes, and tomatoes from the New World. Cassava is a native of South America, but it was introduced into Africa in the sixteenth century and is now a diet staple in many parts of that continent.

Social class is an important element in our food habits. During the first half of this century, the prestige of processed foods was one factor leading to the abandonment of home-grown foods by rural families. The opposite status symbol is now found in some communities, where high-priced, farm-fresh foods are prized by the more affluent class; some Americans are prejudiced against white bread, canned foods, and pasteurized milk. Food habits are also related to our self-image: eating a familiar food associates one with a class or group. Many farm families in the United States partially abandoned their traditional dietary patterns and adopted city diets because they thought theirs was "backwoodsy."

The image of the food itself (a Sunday food, a holiday food, or a summer food, for example) is a factor. The general ideology of the individual, certain ethical concepts such as greediness or frugality, and cultural and personal variations also influence food habits. In the fourteenth and fifteenth centuries the monastic diet was austere and featured frequent fasts and special diets (Henisch, 1976).

Resistance to New Foods

Resistance to new foods, or *neophobia,* is a fact of life: the diet of some peasant societies and of children and adolescents is often incredibly narrow (Siegel, 1957). However, monotony of the dietary regime decreases the overall rated palatability of the diet (Kamen and Peryam, 1961).

New foods challenge the traditional food habits and may upset the dietary balance. Food preferences may start while the child is still breast-feeding. The mother's diet influences the flavor of her milk. In addition, the child is exposed to the cooking odors of the kitchen. Nutritional deficiencies not only depend on economic and cultural conditions but also may be conditioned by surpluses or ready availability of many foods of low nutritional value.

Changing Food Habits

Food habits are constantly changing. The most important factors influencing them at a given time, according to the National Research Council (NRC) Committee on Food Habits (1943, 1945) are: 1) traditional habits and how children learn them; 2) state of food production, processing, and distribution; 3) state of the science of nutrition; 4) methods of disseminating nutritional information; 5) state of medical practice affecting food habits; 6) state of and changes in housing, equipment, and transportation; and 7) state of and trends in child training and education. These reports emphasize how deeply engrained in culture are our food habits and how difficult it often is to change them.

Dependence on the senses leads to a life-style of eating that is more or less independent of nutritional requirements: foods are appreciated for their own sake, "to create the everlasting play between pleasure and dislike" (Renner, 1944). But certainly many of our likes and dislikes are temporary and ephemeral and can be changed. Today's food processors are constantly developing and marketing new food products that tempt the consumer, often with all the fanfare modern advertising provides. However, in general, distrust of what is new or different greatly governs food acceptance. Brand *et al.* (1980) confirm the difficulty of changing food choices and recommend both imitative and innovative approaches to the problem.

Incentives and Techniques for Change. Nowadays many techniques are available for inducing change. Generally when dietary change has become necessary (as during war), the existing food pattern has been more or less retained and the necessary modification made by enrichment, restoration, and nutrification of the common foods and by restricting certain foods (see Chapter 6). The methods used in studying food habits are outlined in the NRC publications mentioned already.

Food habits have changed slowly in Latin America (Adams, 1960). The pre-Conquest diet of corn and beans is still popular among the Indians and poorer classes. The native food production habits persisted until the social structure of the agriculture-producing unit was broken down. In fact, a change in food habits indicated a basic change in the entire social structure and was not due to the developmnent of an interest in better nutrition. Rice, wheat, plantains, cattle, pigs, and chickens have been the most successful new food items for Latin America.

Wittfogel (1960) reports slow changes in food taboos in India, partly due to survival of caste prejudices. On the other hand, marked dietary improvement has occurred in Taiwan as a result of land reform. Glatzel (1961) points out that the increasing urbanization of western Europe (and the United States) has created changes in food habits—fewer garden-grown fruits and vegetables, less work-induced hunger, less home cooking, greater protein need, and less fat need relative to total diet.

In the United States, changes in food habits of immigrants come slowly for Italians but rapidly for Hungarians and Poles. Ethnic food habits are still very important in certain areas in the country. Niehoff (1968) reports that food habits could change in 9 years. Fliegel (1961) has shown that national groups in the country obtain many of their new ideas on food from informal sources, such as friends. Radio and television, women's magazines, and newspapers are all sources of new food ideas, some more effective than others. Dichter (1964) believes that with proper planning food habits can be changed. His plan calls for using advertising that depends on food motivation arising from symbolic or psychological meanings. Just how seriously one can take his suggestions is questionable. Whether consumers have such highly emotional associations with food was not proved by Dichter, nor has experimental proof been offered that this can be accomplished. However, consumer trials of a new food along with an appropriate propaganda campaign could be helpful, especially if it is related to the culture of the community.

Introducing a new product in a developing country is especially risky, according to Cuthbertson (1966), and should be preceded by a close study of the potential market. It is not appropriate to base the publicity for the new food on Western terms such as vitamins and proteins because these have little meaning in many parts of the world.

Value of Prediction. It would be very useful to be able to predict consumer likes and dislikes for new products. Pilgrim and Kamen (1963), on the basis of surveys of American soldiers, reported that about three fourths of the variation in selection of food could be predicted from 1) knowledge of previously expressed food preferences, 2) subjective satiety (or "fillingness of food"), and 3) the amount of fat and protein in the food. Of these, satiety and preference were more important, accounting for about 55% of the variation if considered in combination with fat and protein content, or 66% if considered without.

Role of Instinct

Instinct is not a reliable guide to a balanced food consumption. Many deficiency diseases existed in Europe before there was knowledge of vitamins or other essential elements. Even today, some people have difficulty balancing their diets. The "wisdom of the body" clearly should be supplemented by nutritional information.

Unfortunately, food preferences are not necessarily related to our nutritional requirements. For example, Greenland was colonized about 985 A.D., but within 500 years the colony died out. Study of the skeletons left behind indicates that the settlers may have perished from rickets. Rickets is primarily due to a deficiency of vitamin D, and fish livers are rich in this vitamin. Why didn't the "instinct" of these settlers lead them to eat fish liver, which was plentiful in the area? Nevertheless, early societies learned to avoid poisonous foods, but at how much loss of life we do not know. The Indians of the Americas learned how to remove toxic chemical compounds from raw materials by crushing, extracting, and pressing. (See Heizer and Elasser, 1980, for the California Indian technique of removing toxic materials from acorns.)

Geographical and Cultural Factors

Different nations, social classes, and religious groups have well-developed specific preferences and prejudices for certain foods. Simoons (1976) notes that one may approach food habits holistically, that is, one considers a culture as an integrated whole. Cultural phenomena should be considered in terms of human needs in an active, working society, or one may simply use a geographical or cultural historical approach.

Availability. Food habits develop on the basis of the raw materials available. Beer making predominates in countries where suitable grain is plentiful. Rice is a diet staple in countries where it can be produced economically. The consumption of bananas was originally restricted to tropical countries. Blubber (fat) is eaten by Eskimos because it is more convenient than starchy foods. Meat is consumed to a lesser extent in tropical climates because it is less available and spoils so easily. Of course, as geographical divisions of society developed, these differences in food availability tended to develop as social or group preferences.

The influence of supply on food habits is illustrated by experiences of the

American colonists. They arrived from Europe with their own special food habits. Lacking adequate supplies of European foods, they quickly learned to eat the corn, maple syrup, pumpkins, and dried beans common in the Indian diet. They also learned to use the local fish, game, clams, and oysters (see Todhunter, 1961).

Climate. Climate, too, affects food habits. Grapes do not grow well in a cold climate, nor does wheat in the tropics. Furthermore, a person's own needs change as he or she moves from a cold to a hot climate. In cold climates the motivation to eat is greater and hot foods are preferred, whereas in warm climates, salted foods and water are most often mentioned as preferred for eating and drinking, and people generally cut down on their consumption of meat and fat. However, clothing renders our body environment nearly the same in cold as in warm climates, so today we feel little need for a change in diet, except for a reduction in caloric intake in warmer areas. Even this is not so great a change as commonly believed; in fact, Renner (1944) believes it to be a negligible factor.

The origin of our likes for foods is more important than our dislikes. Many of our likes are deeply rooted in cultural groups, as well as in individuals. Generally, we like or dislike foods according to the society in which we have been brought up. Milk and butter are disliked by the Chinese and Malays; snakes are delicacies to some Australian tribes; fish are unpleasant to many Africans; and "high" game (aged to the point of putrefaction) is favored by some English people. Nevertheless, families may develop and pass on a liking for a food or dish that their neighbors do not like. The Chinese in America have grown their own special kinds of vegetables ever since they arrived. The Chinese food markets are a good example of the continuation of the demand for traditional foods. (See Simoons, 1976, for a general discussion of what he calls our *foodways.*)

Food Taboos

Aside from their origin in availability, food habits have been influenced by taboos of various kinds. The origin of taboos is difficult to fix. Unpleasant and pleasant smells certainly are a probable source; the mystical origin of smells themselves may have led to preferences and dislikes.

The important thing to remember about a taboo is that even though the ostensible reasons for the taboo are imaginary, its force is as great as if they were real. Once a taboo is established, for whatever reason, the adherents avoid the food and may come to abhor it (see also p. 67). Once a food is avoided, it has to be very cheap to attract a consumer again, if at all. In some cases its very cheapness causes consumers to avoid it lest they be thought of as poor!

One source of taboos appears to be religious practice. Le Gros Clark (1968) believes that food taboos developed in the early Neolithic period when priests and medicine men, and they alone, claimed access to the powers that regulated the growth and failure of crops. Religion has continued to have a profound influence on food habits. Simoons (1976) enumerates many religious practices: taboos, fasts, sacredness of the meal, and food supply for the dead and the gods. He notes that India is of the greatest interest in the study of how religion (especially the caste system) influences food habits.

Primitive humans feared the gods and made sacrifices to them. In the Book of Leviticus, the "best" parts of the slaughtered animals—the entrails and blood—were sacrificed to God. Their use was then proscribed to ordinary persons, and gradually prejudices against their use developed. The rite of burnt offering disappeared long ago but some people still do not ordinarily eat the lung, spleen, kidney, or liver. In frontier societies this prejudice often breaks down because of the scarcity of food.

Animal-Food Taboos

Taboos of foods of animal origin have been summarized by Simoons (1961). Some forbidden foods are pork (Muslims, Orthodox Jews, Ethiopian Christians, etc.); beef (Hindus); chickens and eggs (people in parts of India and Africa); horse flesh (Muslims); and dog flesh (widespread).

The use of horse flesh by pagans undoubtedly led to a prejudice against its use by later Christian civilizations. The horse-meat taboo dates from the time of Pope Gregory III in the eighth century. He forbade his German Christian converts horse flesh to show their separateness from the horse flesh-eating pagans. Horse meat is still avoided by many cultures, although it has been eaten in France since the 1880s (Gade, 1975). Similar dietary restrictions are used by several religious minorities to keep themselves separate from the surrounding majority (see also p. 74–75).

The origin of the pork taboo is discussed by Simoons (1961) and by Darby et al. (1977). The traditional explanation is that pork is infested with a parasitic worm (*Trichinella spiralis*) that causes trichinosis in humans (as well as in the rat, pig, and other animals) (see Chapter 4). The taboo appeared in the dynastic period in Egypt (ca. 3500 B.C.), and also much later in the Mosaic and Koranic bans. Domesticated swine were found in Neolithic Egypt. Apparently they were consumed and continued to be consumed by some of the population in certain areas until modern times. Darby et al. (1977) attribute the pork taboo in Egypt to the fact that the pig was the sacred animal of the god Seth-Osiris-Horus. The pork taboo in Egypt, therefore, was primarily maintained by the clergy and nobility.

The folly of attributing prejudices to close association with or dependence on an animal has been discussed by Simoons. The Mongols, who depended so completely on the horse, were horse flesh eaters. The camel-flesh taboo is not easily explainable. Since Muslims eat it, non-Muslims may have sponsored the prejudice as a reaction against the Muslims. This seems to be true of Ethiopian Christians. The dog-flesh taboo may have a variety of origins, but it may be associated with the conflict between pastoral groups and settled, purely agricultural groups. Simoons believes this conflict to be the largest single factor in the origin of taboos. The dog-flesh taboo appears to be irrational because dogs are eaten in Polynesia, Vietnam, China, and certain areas of Africa (Pyke, 1968).

Mosaic Tradition

The development of the Mosaic taboo (which appears first in the Book of Leviticus in the Bible, following the Exodus from Egypt) has been attributed to pork not being produced in regions where the Israelites lived, to a desire by the Israelites to set themselves apart from their pig-eating neighbors, to revelation, to codification by the clergy, and to other reasons, including health. Pork avoidance may be due to the fact that pigs are unsuited to nomadic living. Simoons (1961) suggests that invasions of pastoral cultures brought along with them the prejudice against swine. He dismisses the idea that the disease trichinosis or the pigs' scavenging nature (p. 73) has anything to do with the pork taboo. The prohibition of the Jews against pork and rabbit meat also may have been economic: they were a pastoral people and naturally wanted to get the most for their cattle, sheep, and goats.

In Judaism, animals are clean or unclean. The former include those that chew their cud and whose hooves are divided. The pig and camel do not satisfy both requirements, though they satisfy one of them. Only fish with both fins and scales are allowed to be used as food. Blood is sacred and therefore taboo. The internal fat of an animal is also taboo. Meat and dairy foods are not to be eaten together. Thus in a truly Orthodox Jewish home two sets of cooking and serving utensils must be maintained—one for meat and the other for dairy foods. The Orthodox Jew also waits 6 hours after consuming meat before consuming milk or dairy foods, but meat may be consumed one hour after drinking milk.

On Jewish holy days and festivals, foods play important symbolic roles. During the 8-day celebration of Passover, leavened bread cannot be eaten, but foods having specific symbolic meanings are eaten. A roasted egg symbolizes the burnt offerings made in the Temple at Jerusalem; maror (a bitter herb such as horseradish) recalls the bitterness of slavery; and karpas (usually parsley or celery) represents the poor food fed the Israelites during their years of slavery.

The death penalty was once required of Jews who ate leavened bread during Passover. Though ostensibly used as a memorial to the flight from Egypt (Jacob, 1944), it is believed to be a true taboo. The high priest of Jupiter in Rome was also prohibited from eating leavened bread. The present concept is that this was a sin offering and that the god should have only unspoiled food.

Muslim Tradition

Avoidance of pork by Muslims is believed to have been due to Mohammed's desire to distinguish his followers from the pork-eating Christians, their chief rivals. (In the Pacific Islands, on the contrary, pig flesh was, and still is in certain areas, a prestige food and was given to women only as a special favor!) A number of dietary regulations are prescribed by the Koran or by Muslim scholars. The fast of Ramadan lasts from sunrise to sunset for a lunar month. Not even a sip of water is permitted during the day. A special type of leavened bread is eaten, and meals are generally light. Muslims are forbidden wine or other intoxicating beverages. A special ritual is followed in killing animals, otherwise they may not be eaten. Some of the Jewish and Muslim practices are similar in character.

Indian Tradition

Simoons (1961) finds the avoidance of beef in India is not easily explained, but he reluctantly concludes that the most likely hypothesis is that it is due to belief in the sacred character of cattle (i.e., belief in the sanctity of life). No satisfactory explanation of the taboo against chickens or eggs has been put forward either. In some castes, onions, turnips, lentils, mushrooms, and coconuts are also not eaten. By contrast, Buddhism has few food taboos. In theory, Buddhists are supposed to be vegetarian, but many are not.

The special problems of India and of other countries where the principle of *ahimsa* (nonviolence) prevails must be considered (Simoons, 1976). Even in the West vegetarianism is not unknown, particularly among some religious groups. There are over 2 million vegetarians in the United States. The basic problem of such diets is the possible deficiencies in protein and vitamin B_{12} that may result, particularly (as in India) when caloric intake is already limited. Efforts to improve the protein intake by greater use of vegetable proteins, milk, eggs, and fish have been instituted in India and elsewhere for this reason. The assumption is that it is easier to accept the *ahisma* principle than to change it.

Other Food Traditions

Fast days, as a sign of devotion or penance, are widespread. On certain feast days, special foods are prepared, such as cakes and eggs at Easter and special boiled wheat grains (*koliud*) following the death of a loved one (see Lowenberg *et al.,* 1968, for other examples).

Magic as a factor in food prejudices is discussed by Pyke (1968). The hippomanes of the foal, a fleshy excrescence, was believed by the Romans to be especially potent. The mandrake (*Mandragora officinarum*) fruit was thought to have magical properties (Genesis 30:14–16). In the Middle Ages people ate long-lived plants hoping to prolong life. Walnuts were eaten for brain disease because the outer hull superficially resembles the exposed brain.

Curious food customs persist. In Japan, entomophagy (eating of insects) may be observed. The Masai tribe in Kenya drinks cattle blood. Insects are not commonly used for food in Europe, but they have been so used in Australia, Africa, Asia, and the Americas, according to Bodenheimer (1951). Perhaps it is because insects are high in protein, which is in short supply in the tropics. Mushroom preference (mycophagy) is common among the Northern Slavs (Wasson and Wasson, 1957). Mushrooms are used by some native Americans for their hallucinogenic properties.

Many taboos extend to mixtures with other foods, even to mixtures in small amounts. Taboos are also applied to sexes, age groups, and status groups. On the other hand, all societies have some members who do not conform to the taboo. The origin of other taboos is more complex: rule of elders; fear of disease; fear of infertility; totemic, dualistic concepts; attempts to conform to conventional or popular concepts; or male secret societies.

Taboos of whatever origin often tend to gradually erode away. The sense of adventure, new cultural patterns, mixtures of cultures in business, and association of different ethnic and cultural groups in elementary schools or in social groups are common factors influencing this breakdown. Because the great migrations tended to dissipate religious and cultural prejudices, it is remarkable that the Jewish people have retained their food habits for so long. However, adherence to their various dietary customs has become less strict for some Jewish people in the United States.

Preferences and Avoidance

Genetic Factors

Simoons (1976) has noted a possible genetic origin of certain food avoidance. Lactose malabsorption is one example. Milk is the universal food of newborn mammals, but after 2 to 4 years of age, most humans lack the enzyme lactase that hydrolyzes lactose (milk sugar). Simoons believes that in periods of nutritional stress, aberrant children with some lactase-producing ability survived by drinking milk. Thus a lactase-producing ability persists beyond infancy. Adults of northern Europe and most white American ethnic groups produce lactase in their digestive tract and therefore can tolerate milk. Continuous feeding of milk does seem to increase tolerance to lactose. Genetic differences between ethnic groups are important, however: many black Americans are milk-intolerant from lack of lactase. (See Fig. 23 for the ethnic distribution of lactose intolerance.)

The present hypothesis is that dairying did not appear until about 10,000 years ago, long after the mammalian pattern of lactase production had been well established in humans, which was 100,000 years ago. Chance mutation, providing lactase-production genes, may have occurred in some humans; milk-drinking humans had an advantage over lactose-intolerant individuals. Large-scale distribution of milk powder to lactose-intolerant populations

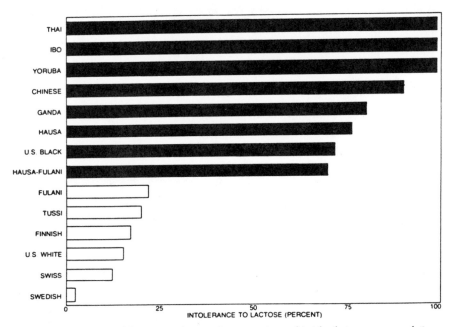

Fig. 23. Distribution of lactose intolerance by percentage of total ethnic group population. Intolerance varies widely among populations. The bars are based on tests conducted by a number of investigators by different methods; they may not be strictly comparable or accurately reflect the situation in entire populations. Among the groups studied to date, lactose intolerance is prevalent except among northern Europeans (and their descendants) and herders in Africa. (From Kretchmer, 1972. Copyright © 1972 by Scientific American, Inc. All rights reserved.)

should probably be modified. Fortunately, lactose-free milk powder can be manufactured and is available.

Cultural Factors

Regional preferences for particular foods still exist in the United States, even among transplanted college students. Einstein and Hornstein (1970) found that grits, black-eyed peas, lima beans, and iced tea were preferred only in the South. Chili (and to a lesser extent roast leg of lamb) were preferred in the West; clam chowder was preferred in the Northeast. In the region near Philadelphia, hot roast pork sandwiches, roast pork, and grilled pork chops were all disliked by some.

The diet of contemporary urban Americans differs from that of small, non-industrialized, nonaffluent, and relatively homogeneous societies and communities (Jerome, 1975). Because of the wide and stable variety of attractive foods available, Americans have become very selective. Promotional programs (both for food and nutrition) influence food selection and use. In general, American societal goals emphasize the benefits of food variation, self-fulfillment, and achievement.

The typical Anglo-American diet is not very characteristic; the cuisine of many societies is very specific. The advantage of this is that our experience as to what is nutritious or poisonous can be passed along. A cultural cuisine consists of 1) primary (staple) and secondary foods that provide most of the calories and nutrients, 2) foods prepared in specific ways, and 3) characteristic use of flavorings. There are also cultural rules about how the foods are consumed (Rozin, 1976). Rozin noted that some cuisines can best be explained as an art form. Religious and social factors combined with a cultural cuisine provide social groups with a specific identity. Some cuisines change very slowly. The Iranian peasant cuisine is essentially the same today as it was 6000 to 7000 years ago.

Spices are one of the characteristic features of many cuisines, although they contribute little to nutrition in the amounts used. Generally they are used in such small amounts that they provide few calories, protein, vitamins, or minerals. Their stimulation of gastric secretion (black pepper) or their bacteriostatic effect is not considered important. Even the concept that they add variety to the diet is subject to some doubt, considering how monotonously some are used (e.g., curry in India and chili in Mexico). Chili pepper is probably the most frequently used flavoring in nonindustrial societies, but it is also widely used in some industrial societies, like Texas (Moore, 1970).

World trade and travel also reduce the tendency to maintain strong local food preferences. Trade has been especially important in introducing exotic and common foods to distant lands. The Portuguese fishermen catch their cod off Newfoundland, and cod has become a stable in the diet of the people of Portugal and Brazil.

To summarize, food habits and taboos originated early in human development. They arose from economic and cultural conflicts of nomadic and settled peoples, from religious customs, availability of foods, climate, commerce, and invention. At present we acquire them from our parents and modify them more or less as we grow older because of various external factors: advertising, class or status symbolism, or experience. The so-called dietary instinct is mainly a distrust of the unknown and is an unreliable nutritional guide.

Food Quality—Attributes and Examples

Based on extensive study and observation of the American consumers over the past few decades, we have learned much about what they believe to be the important criteria of food quality. To some extent we have also learned how to identify and measure quality attributes. In this section we shall examine this subject in a general way, leaving for later a more detailed treatment of attribute measurement and the way quality is controlled in the food industry (see Chapters 7 and 8).

Food scientists generally agree that there are at least six important attributes of food: safety, purity, convenience in use, shelf life, functional performance, and nutritional value. Let us now define each of these terms and provide some examples. Students should recognize them as the most important factors in determining the quality of the food that consumers seek in the food market.

Safety

Safety is the term food scientists use to denote that all foods should be free of health hazards, especially toxic chemicals and infective agents that cause illness. For example, in North America it is legally required that beverage milk products sold to consumers must be free of disease organisms such as those causing tuberculosis, undulant fever, and streptococcal sore throat. Similarly, all meat and poultry products must contain only very minimal levels of the antibiotics, hormones, and toxic chemicals used in the production of livestock and poultry. Likewise, other foods must contain no more than insignificant residues of the toxic chemicals used in agriculture and in industry that sometimes find their way into food (see Chapter 9 for details).

The food scientist must be familiar and able to cope with the numerous complex issues involved in assuring the safety of food. We will examine in detail the safety problems in foods in Chapter 4 and to some extent in

Chapters 7, 8, and 9, which deal with food processing and preservation and food laws and regulations.

Purity

Purity is an important attribute of food, although it does not rate as high a priority as safety in the minds of most American consumers. Because purity is sometimes confused with safety, even by some food professionals, we will make a special effort to distinguish clearly between these two terms.

Purity concerns those materials sometimes found in food that are usually esthetically unacceptable to American consumers. There are two types of these materials: 1) unwanted and unacceptable portions of the raw materials used for converting into processed food and 2) foreign objects from the processing environment that may accidentally contaminate the food. Purity *does not* concern those materials or agents that are hazardous to health, which we have just described under safety.

Consider the following examples. First—a simple case—consumers expect tomato juice to be free of the seeds and skins. These are not harmful, but consumers just do not want them in tomato juice; fortunately, this is readily accomplished commercially. A slightly more complicated example is that homemakers expect a ready-to-cook turkey to be free of unsightly pinfeathers. Again, it is not unhealthy to consume these parts of a turkey, but we just do not like the idea of doing so. Commercially, it is not always easy or economical to remove all pinfeathers, completely, especially from dark-feathered poultry. So although homemakers might prefer a pinfeather-free turkey, they do not necessarily refuse one with a few pinfeathers showing, especially if it is offered at a bargain price. The customer knows that eating a few pinfeathers is not going to make anyone ill and, in any case, pinfeathers can usually be removed or concealed by the cook.

Esthetics. A different kind of purity problem is that involved with esthetically objectionable but not unhealthful contaminants that sometimes get into food. A good example is the infestation of flour with weevils (a beetle commonly found in cereal products). Homemakers do not expect or want to find weevils in flour, but most of them know that occasionally weevils will appear in the flour they purchase. Most homemakers know that there is no danger to the health of their families from eating a few live weevils, but they just do not like the idea.

There are numerous other examples of purity problems in food. Although these problems are basically less important than those of safety, because of their nonacceptance by consumers they can become a major preoccupation for a food scientist working in quality-control programs. We shall encounter some of these problems in Chapters 7, 8, and 9.

Sensory Properties

Sensory properties are defined as those characteristics of a food that are detected by the sensory organs. Some common terms expressing these sensory sensations are *appearance/color, aroma/bouquet, flavor,* and *texture.* Here we are dealing with a wide range of important characteristics that largely determine the acceptability of food by most consumers.

Study of the sensory evaluation of foods is just now beginning to receive the attention from scientists that it deserves. It has also gradually gained the understanding and respect of food scientists. Because of its great importance in food science and technology, the basic principles of sensory evaluation and their application to food processing and preservation will be discussed in detail in Chapter 5 and also to some degree in Chapters 7 and 8, which deal with processing and preservation.

Convenience

Convenience in use is practically self-explanatory: ease in getting the food out of the package and preparing it for use. It is an attribute of food and its package that has gained enormously in importance as our society has become urbanized, especially as husbands and wives both have become employed outside the home and want to make food preparation less of a chore. Convenience in use has also become important in institutional and restaurant foods because of rising labor costs and a demand for a greater variety of foods to be served in these places. Many food items and even whole meals are available precooked or partially cooked. We shall be discussing this attribute again in Chapters 7 and 8.

Shelf Life

Shelf life concerns the ability of foods to maintain their overall quality, especially their safety, sensory properties, and nutritional value, during distribution and marketing and while the food is in the consumers' hands. Assuring an appropriate shelf life for a given food is a major preoccupation for the food scientist in industry. It will be discussed in some detail in Chapters 7 and 8.

Functional Properties

Functional properties are those characteristics of food, especially of ingredients, that perform in certain essential ways during processing and preservation. Examples include the foaming and whipping power of egg white and cream; the ability of wheat flour along with other ingredients to form an elastic dough that, during fermentation, preparation, and baking, is transformed into bread; the thickening and gelling properties of starch; and the emulsifying power of egg yolk to convert vinegar and oil, with proper physical manipulation, into a creamy, smooth emulsion (mayonnaise). These are important characteristics in the manufacture of many common foods (see Figs. 24 and 25). This attribute will be discussed further in Chapters 7 and 8.

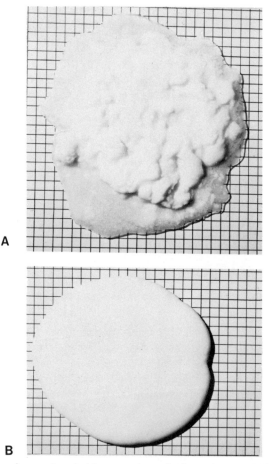

Fig. 24. Functional properties of white sauces frozen for 5 months at −18°C (0°F), then thawed. (A) Sauce made with wheat flour; (B) sauce made with waxy rice flour. (From Hanson *et al.*, 1951. Copyright © by Institute of Food Technologists.)

A

B

Fig. 25. Salad dressings stored for 3 months at −18°C (0°F), then thawed. (A) Containing peanut oil; (B) containing cottonseed oil. (From Hanson and Fletcher, 1961. Copyright © by Institute of Food Technologists.)

Nutritional Value

Nutritional value is self-explanatory. Food scientists in recent years have had to be much more conscious of and concerned about nutritional values, especially as consumers and government officials have shown much greater interest and concern about the nutritive value of our food supply. Fortunately we now know considerably more about the effects of processing and preservation technologies on the nutrient content of foods, ways of maximizing it, and even how to enhance the nutrient content through fortification.

Nutrition and food science will be discussed in detail in Chapter 6 and to some extent in Chapters 7 and 8.

REFERENCES

Adams, R. N. (1960). Food habits in Latin America: a preliminary historical survey. *In* "Human Nutrition; Historic and Scientific" (I. Goldston, ed.), Vol. XVII, pp. 1–22. International Univ. Press, New York.

Bodenheimer, F. S. (1951). "Insects as Human Food." Junk, The Hague.

Brand, J. G., Kare, M. R., and Naim, M. (1980). *In* "Nutrition, Food and Man. An Interdisciplinary Perspective" (P. B. Pearson and J. R. Greenwell, eds.), p. 105. Univ. of Arizona Press, Tucson.

Cuthbertson, W. F. J. (1966). Problems in introducing new foods to developing areas. *Food Technol. (Chicago)* **20**, 66–68.

Darby, W. J., Ghalioungui, P., and Grivetti, L. (1977). "Food: The Gift of Osiris," Vols. 1 & 2. Academic Press, New York.

Dichter, E. (1964). "Handbook of Consumer Motivation: The Psychology of the World Objects." McGraw-Hill, New York.

Einstein, M. A., and Hornstein, I. (1970). Food preferences of college students and nutritional implications. *J. Food Sci.* **35**, 429–436.

Fliegel, F. C. (1961). Food habits and national backgrounds. *Bull. Pa. Agric. Exp. Stn. No. 684*, 1–24.

Gade, D. W. (1975). Horsemeat as human food in France. *Ecol. Food Nutr.* **5**(1), 1–11.

Garcia, J., and Hankins, W. G. (1975). The evolution of bitter and the acquisition of toxiphobia. *Olfaction Taste, Proc. Int. Symp., 5th, Univ. Melbourne, 1974* pp. 1–126.

Glatzel, H. (1961). Entwicklungstendenzen in der Kostwahl und ihre Auswirkungen auf die Ernährungssituation. *Ernaehr.-Umsch.* **8**, 231–233.

Hanson, H. L., and Fletcher, L. R. (1961). Salad dressings stable to frozen storage. *Food Technol. (Chicago)* **15**, 256–262.

Hanson, H. L., Campbell, A., and Lineweaver, H. (1951). Preparation of stable frozen sauces and gravies. *Food Technol. (Chicago)* **5**, 432–440.

Heizer, R. F., and Elasseer, A. B. (1980). "The Natural World of the California Indians." Univ. of California Press, Berkeley.

Henisch, B. A. (1976). "Fast and Feast: Food in Medieval Society." Pennsylvania State Univ. Press, University Park.

Jacob, H. E. (1944). "Six Thousand Years of Bread. . . ." Doubleday, New York.

Jerome, N. W. (1975). On determining food patterns of urban dwellers in contemporary United States society. *In* "Gastronomy. The Anthropology of Food and Food Habits" (M. L. Arnott, ed.), pp. 91–111. Mouton, The Hague.

Kamen, J. M., and Peryam, D. R. (1961). Acceptability of repetitive diets. *Food Technol. (Chicago)* **15**, 173–177.

Kretchmer, N. (1972). Lactose and lactase. *Sci. Am.* **227**(10), 70–78.

Le Gros Clark, F. (1968). Food habits as a practical nutritional problem. *World Rev. Nutr. Diet.* **9**, 56–84.

Lowenberg, M. E., Todhunter, E. N., Wilson, E. D., Feeney, M. C., and Savaga, J. R. (1968). "Food and Man." Wiley, New York.

Moore, F. W. (1970). Food habits in non-industrial societies. *In* "Dimensions of Nutrition" (J. Du Pont, ed.), pp. 181–221. Colorado Assoc. Univ. Press, Boulder.

National Research Council's Committee on Food Habits (1943). The problem of changing food habits. *Bull. Natl. Res. Counc. (U.S.)* **108**, 1–177.

National Research Council's Committee on Food Habits (1945). Manual for the study of food habits. *Bull. Natl. Res. Counc. (U.S.)* **111**, 1–142.

Niehoff, A. (1968). Food habits and cultural patterns. *In* "Food, Science and Society," p. 54–68. Nutr. Found., New York.

Pilgrim, F. J. (1957a). The components of food acceptance and their measurement. *J. Clin. Nutr.* **5**, 171–175.

Pilgrim, F. J. (1957b). The components of food acceptance and their measurement. *Nutr. Symp. Ser.* No. 14, pp. 69–73.

Pilgrim, F. J., and Kamen, J. M. (1963). Prediction of human food consumption. *Science* **139**, 501–502.

Pyke, M. (1968), "Food and Society." Murray, London.

Renner, H. D. (1944). "The Origin of Food Habits." Faber & Faber, London.

Rozin, P. (1976). Psychobiological and cultural determinants of food choice. *Life Sci. Res. Rep.* **1**, 285–312.

Seward, J. P., and Greathouse, S. R. (1973). Appetitive and aversive conditioning in thiamine deficient rats. *J. Comp. Physiol. Psychol.* **83**, 157–167.

Siegel, P. S. (1957). The repetitive element in the diet. *Nutr. Symp. Ser.* No. 14, pp. 60–62.

Simoons, F. J. (1961). "Eat Not This Flesh." Univ. of Wisconsin Press, Madison.

Simoons, F. J. (1976). Food habits as influenced by human culture: approaches in anthropology and geography. *Life Sci. Res. Rep.* **1**, 313–329.

Todhunter, E. N. (1961). The history of food patterns in the U.S.A. *Proc. Int. Congr. Diet.* **3**, 13–15.

Todhunter, E. N. (1979). Food habits, food faddism and nutrition. *In* "Nutrition and the World Food Problem" (M. Rechcigl, Jr., ed.), pp. 267–294. Karger, Basel.

Wasson, V., and Wasson, R. G. (1957). "Mushrooms, Russia and History." Pantheon Books, New York.

Wittfogel, K. A. (1960). Food and society in China and India. *In* "Human Nutrition; Historic and Scientific" (I. Goldston, ed.), Vol. 17, pp. 61–77. International Univ. Press, New York.

Yudkin, J., and McKenzie, J. C., eds. (1964). "Changing Food Habits." Macgibbon & Kee, London.

SELECTED READINGS

Amerine, M. A., Pangborn, R. M., and Roessler, E. B. (1965). "Principles of Sensory Evaluation of Food." Academic Press, New York.

Blakeslee, A. F., and Salmon, T. N. (1935). Genetics of sensory thresholds—individual taste reactions for different substances. *Proc. Natl. Acad. Sci. U.S.A.* **21**, 84–90.

Birch, G. G., Brennan, J. G., and Parker, K. J., eds. (1977). "Sensory Properties of Foods." Appl. Sci., London.

Dallenback, K. M. (1939). Smell, taste and somesthesis. *In* "Introduction to Psychology" (E. G. Boring, H. S. Langfeld, and H. P. Weld, eds.), pp. 600–626. Wiley, New York.

Kramer, A., and Twigg, B. A. (1966). "Fundamentals of Quality Control for the Food Industry." Avi, Westport, Connecticut.

National Academy of Sciences (1976). "Objective Methods for Food Evaluation," pp. 7–27, 145–153, 207–213. Natl. Acad. Sci., Washington, D.C.

Pfaffmann, C. (1951). Taste and smell. *In* "Handbook of Experimental Psychology" (S. S. Stevens, ed.), pp. 1143–1171. Wiley, New York.

Raunhardt, O. and Escher, F. eds. (1977). "Sensory Evaluation of Food." Forster-Verlag, Zurich.

Stellar, E. (1976). The CNS and appetite: historical introduction. *Life* Sci. Res. Rep. **1**, 15–20.

Tsai, L. S. (1976). Quality evaluation of eggs and poultry meats. *In* "Objective Methods for Food Evaluation: Proceedings of a Symposium, November 7–8, 1974," pp. 79–92. Natl. Acad. Sci., Washington, D.C.

Weiffenbach, J. M., Daniel, P. A., and Cowart, B. J. (1980). Saltiness in developmental perspective. *In* "Biological and Behavioral Aspects of Salt Intake" (M. R. Kare, M. J. Fregly, and R. A. Bernard, eds.), pp. 13–29. Academic Press, New York.

Chapter 4

Food Safety and Principles for Its Control

Most food scientists, as well as laymen, agree that safety is of paramount importance among the quality attributes of food. In spite of the public outcries and the controversies that currently surround the subject, few countries, if any, have a better record of food safety than the United States (see Chapter 11). This is so because of the large amount of research and development that has been done on the subject over the past few decades and to the extensive educational efforts that have been made to disseminate research findings among those working with or distributing foods. In spite of this fine record, there still remain associated with food health hazards about which we have insufficient knowledge. Scientific inquiry must be and is being actively pursued on this important subject here and in many other countries around the world. The food scientist must understand the nature of these hazards and learn how to deal with them effectively, especially if he or she is working in industry or for a food-related regulatory agency. In this chapter we shall examine a few of the more important facets of food safety and means for their control.

Health hazards associated with food may be divided into two distinct types: 1) infections and 2) intoxications. *Infections* are diseases that require the proliferation of living organisms and their invasion of body tissues to produce their adverse effects. *Intoxications* are diseases caused by the ingestion of toxic chemicals that adversely affect the body. It is important to recognize this distinction because it frequently dictates which preventive or corrective measures must be taken to control the hazard. We will examine a few of the more important food-borne infections and intoxications.

Food-Borne Intoxications and Infections

Intoxications—Naturally Occurring Toxins

Highly Toxic Chemicals. Too many people who should know better forget that many products of nature are not safe to eat; in fact, some are highly poisonous. For example, certain types of wild mushrooms are so toxic as to cause death within a few hours of consumption. Every year some of our citizens die from gathering and eating such plants. Another less well-known example is the puffer fish, the consumption of which causes a number of deaths each year in Japan. Fortunately this fish is not common in the United States and in any case is not considered to be edible here.

In addition to those plants and animals that contain highly poisonous compounds as normal constituents of their tissues, there are others that acquire toxic substances from the environment in which they live. A notable example is certain shellfish (especially clams and mussels) that may contain a highly toxic chemical. The problem occurs only during certain seasons of the year (spring and summer) and for this reason the gathering of these shellfish for food is illegal on the West Coast of the United States during late May and the summer months. Persons consuming the affected shellfish usually suffer respiratory and cardiovascular failure; sudden death following consumption is not uncommon.

Less Toxic Chemicals. Less toxic substances are also found in some common food raw materials. Table 12 gives their common names, chemical makeup, plant sources, and toxicity symptoms. Fortunately, many of these can be so processed as to remove or destroy the toxins (e.g., cassava, pulses, fava beans, chick-peas, etc.); however, others cannot be so processed.

Nutritional Inhibitors. Chemicals found in certain food plants and animals interfere with the utilization of essential nutrients in the body. Although not as dangerous as the poisons just discussed, they must be recognized as hazards to human health. Table 13 presents a list of some common nutritional inhibitors, their sources, chemical nature, health effects, and means for counteraction. It is of special interest to the food scientist to note that, fortunately, heat treatment or the addition of supplementary nutrients are effective countermeasures. Both can be accomplished during food processing and preservation operations.

TABLE 12
Some Naturally Occurring Toxins in Food Raw Material

Toxin	Chemical nature	Main food sources	Major toxicity symptoms
Protease inhibitors	Proteins	Beans (soy, mung, kidney, navy, lima); chick-pea; peas; potato (sweet and white); cereals	Impaired growth and food utilization; pancreatic hypertrophy
Hemagglutinins	Proteins	Beans (castor, soy, kidney, black, yellowjack); lentils; peas	Impaired growth and food utilization
Cyanogens	Cyanogenic glucosides	Peas and beans; pulses; flax; fruit kernels; cassava	Cyanide poisoning
Gossypol pigments	Gossypol	Cottonseed	Liver damage; hemorrhage; edema
Lathyrogens	β-aminopropio-nitrile and derivatives	Chick-pea	Osteolathyrism (skeletal deformities)
	β-N-oxalyl-L-α,β-diaminopropionic acid	Chick-pea	Neurolathyrism (damage to central nervous sytem)
Cycasin	Methylazoxymeth-anol	Nuts of *Cycas* genus	Cancer of liver and other organs
Bracken fern carcinogen	Unknown	Young fronds of fern	Cancer of intestinal tract and other organs
Favism	Unknown	Fava beans	Acute hemolytic anemia

TABLE 13
Nutritional Inhibitors Found in Foods

Sources	Compound	Physiological action	Counteraction
Cabbage	Thioglycerides	Hypothyroidism	Additional iodine
Cereals	Phytin	Binds calcium	Additional calcium
Corn	Unknown	Decreases effectiveness of niacin	Additional niacin
Egg white	Avidin	Binds biotin	Heat inactivation
	Conalbumin	Binds iron	Heat inactivation
Fish	Thiaminase	Destroys thiamin	Heat inactivation
Rhubarb	Oxalic acid	Binds calcium	Additional calcium
Soybeans	Unknown	Enlarges thyroid	Additional iodine
	Lipoxidase	Destroys vitamin A	Additional vitamin A
Spinach	Oxalic acid	Binds calcium	Additional calcium

Allergenic Compounds. There are chemicals (most appear to be proteins) that, if consumed by certain people, lead to skin rash, inflammation of the gastrointenstinal and respiratory tracts, and other adverse effects. Symptoms may occur within a few moments or a day or two after consumption of the responsible food. Most of us are aware of this kind of hazard from personal experience or from that of friends or relatives who suffer food allergies. A food scientist must keep this hazard clearly in mind, especially if he or she is involved in developing special dietary foods or in the design of labels for packaged foods. It is important for persons with any of these allergies to know the components of the foods they buy in order to avoid allergic reactions. Some common allergenic foods are strawberries, eggs, milk, cereals, nuts, fish, and shellfish.

There are other plant and animal foods that contain toxic compounds, but the foregoing examples should suffice to alert us to the potential hazards that exist in consuming certain items from nature's bounty! The food scientist certainly must keep this in mind, especially when involved in the development of new food products from exotic plants and animals that do not have a proven history of being safe raw materials for food. In addition, it is to be remembered that many members of the human race are allergic to certain foods.

Intoxications—Chemical Residues

It is becoming increasingly clear that food raw materials can and sometimes do become contaminated with chemicals (or their derivatives) used in agriculture and/or by industry. Some of current concern are the following.

Agricultural Chemicals. A wide variety of compounds are used in the production of food crops and animals: 1) for plant production—fertilizers, pesticides, herbicides, growth regulators, and others; 2) for livestock and poultry production—drugs, antibiotics, growth stimulants, hormones, pesticides, and others. Although strict government regulations govern the use of these chemicals in America (see Chapter 9), some residues seem inevitably to find their way into the harvested crops and into the tissues of food animals and food products obtained from them, such as milk and eggs. Many of these compounds are harmless to humans but many others are toxic, some highly so (e.g., parathion and related pesticides). Symptoms of intoxication vary widely depending on the type of compound and the amount consumed. Most damaging are those that are carcinogenic, mutagenic, or teratogenic.

Government regulations stipulate the amount, if any, of the residue of an agricultural chemical that is allowed in foods sold for human consumption. It is the job of food scientists, especially those in industry and regulatory agencies, to see to it that the tolerance levels permitted in processed foods are not exceeded. This subject, including preventive measures, is given further treatment in Chapters 7, 8, and 9.

Industrial Chemicals. A great variety of chemicals is used by industry in highly industrialized countries such as the United States. In fact, they are too numerous to list here. Many are harmless, but others are highly toxic to humans (e.g., mercury and arsenic compounds). The food scientist and public health personnel must work together to detect their possible presence, then reject contaminated raw material or so process it as to avoid producing finished products that contain illegal amounts of dangerous contaminants. More on this topic will be found in Chapters 7, 8, and 9.

Food Processing Chemicals. Some of the compounds used by the food industry itself are toxic and, unless great care is exercised, residues can persist in finished products. The following compounds are among those that have hazardous residues: 1) pesticides (e.g., methyl bromide and other fumigants); 2) detergents and sanitizers (e.g., chlorine- and iodine-containing compounds); and 3) bleaching and maturation agents such as organic chlorine compounds and hydrogen peroxide. Government regulations control the use of all chemicals in foods and govern the amount, if any, of residues treated foods may contain. The role of the food scientist in monitoring and controlling these deliberately used chemicals in processing and preservation and in controlling residue levels within legal limits is all-important (see Chapter 9).

Intoxications—Toxins of Microbial Origin

A number of microorganisms that are associated with foods produce toxins in ingredients or in finished products. A few of the most important types encountered in the United States will be discussed. Preventive measures are described in Chapters 7 and 8.

Staphylococcal Enterotoxicoses. The disease caused by toxins produced by strains of *Staphylococcus aureus* is characterized by nausea, vomiting, abdominal pain, and diarrhea, and is commonly referred to as *food poisoning*. Symptoms usually occur within a few hours of consuming contaminated food. Products commonly associated with the disease include custard-filled pastries, egg- and milk-based foods, and certain meat dishes. These are some of the most common and serious food-borne diseases in America.

Botulism. The disease known as botulism is caused by potent toxins produced by strains of the anaerobic bacterium *Clostridium botulinum*. Symptoms include nausea, vomiting, and diarrhea, followed by blurred vision and other ophthalmic changes, and by general muscular weakness. Death is not infrequent. The foods most frequently associated with botulism are the nonacid canned foods, especially canned vegetables, canned mushrooms, canned fish, smoked fish, and certain types of canned soup. Botulism is not common in North America. However, a few cases are reported every year from the consumption of home-canned vegetables and sometimes from commercially processed foods such as canned tuna, canned soups, and canned mushrooms. This problem is a serious concern to food scientists, and much research is being carried out to eliminate the disease, especially in commercially processed products.

Bacillus Aureus Gastroenteritis. Another intoxication associated with food is caused by the aerobic bacterium *B. aureus.* The symptoms are similar to those of the staphylococcal enterotoxicoses. Foods associated with this disease vary widely, from chicken dishes to certain puddings. Much more research is needed before we will know how important it really is among the food-borne intoxications of microbial origin in this country.

Mycotoxicoses. Other diseases are caused by toxins produced by certain molds. Though a few (e.g., ergotism from mold-infected overwintered grains) have long been known to affect humans, evidence thus far is insufficient to implicate the many other fungal toxins that are known to affect a number of animal species. Symptoms of the intoxication in experimental animals, livestock, poultry, and fish vary widely, but at least one of the toxins, aflatoxin, causes cancer in fish. Mycotoxicoses are most commonly associated with cereal grains, nuts, oilseeds, and products manufactured from them. This disease warrants and is receiving a greatly expanded research effort, especially with respect to its possible importance as a food-borne intoxication affecting humans. From accumulating evidence, it seems likely that the mycotoxins will take a place as an important and serious food-borne malady of humans in certain parts of the world.

Bacterial and Viral Infections

As already indicated, infectious diseases require the proliferation of the causative organisms and their invasion of the tissues to produce their characteristic symptoms. Some of the more important food-borne infections of bacterial and viral origin will be discussed next; preventive measures are discussed later on and in Chapters 7 and 8.

Salmonelloses. Some gastrointestinal diseases are produced by *Salmonella* bacteria. The symptoms are quite similar to the intoxications caused by staphylococci and bacilli mentioned in the previous section, and the term food poisoning is applied to all these diseases, even though one group consists of intoxications and the other is infectious. (An infectious disease can be contracted by a person directly from another person who is carrying the infection.) Salmonelloses are commonly associated with eggs, poultry, veal and other meats, milk, and food products made from them.

Clostridium perfringins Gastroenteritis. Still another food-borne infection is caused by the anaerobic bacterium *Cl. perfringens.* Again, the symptoms are very similar to those of the other gastrointestinal diseases with which it is frequently confused. Although much of the work on this disease has been carried out in the United Kingdom, there seems to be little doubt that the organism is widespread and that the disease is probably common in North America. Foods associated with this disease include raw or undercooked beef, lamb, pork, poultry, and fish, and food products made from them.

Vibro Gastroenteritis. This infection, widely occurring in Japan, is caused by the consumption of contaminated marine fish and products made from them. The responsible organism is a salt-tolerant facultative anaerobic bacterium. The symptoms of the disease are similar to the others affecting the gastrointestinal tract. There is some doubt among experts whether the disease occurs in this country, although a recent epidemic originating in the Far East spread elsewhere into areas of the world with a warm climate. More research is needed on this infection, especially regarding its possible presence in this country.

Viral Infections. Viral agents must not be overlooked as a cause of health hazards in food. Although only a limited amount of research has been done on viral infections from contaminated food, at least two viruses have been so implicated. An adenovirus was recently claimed to be the cause of a gastrointestinal incident in humans and an as yet unidentified viral agent has been found to cause hepatitis in humans (Riemann and Bryan, 1980). The latter disease is characterized by inflammation of the liver, fever, jaundice, and gastrointestinal upset. The foods with which these viral infections have been associated include uncooked shellfish, salads, and certain sandwich items.

Parasitic Infections

Many food animals are infested with parasites that, when present in food produced from them, cause disease in humans. There is a variety of such parasites in wild and domestic species, but apparently only one or two of them are responsible for health hazards in North America. One important type will be discussed here. Prevention of these infections is discussed in Chapter 8.

Trichinella spirala infection. In humans the disease known as trichinosis is the best known and understood. The parasite has a life cycle that involves swine, certain other carnivores, and sometimes humans. Symptoms include gastrointestinal disturbances, muscle pains, and fever. At times, the infection in humans is so mild as not to produce recognizable symptoms and therefore it goes undetected. Pork from noninspected slaughter plants is the most common food involved in this ailment in America. Undercooking of such pork is the common cause of viable agents remaining in the meat.

Principles for the Control of Food Safety

In our country safety of the food supply is considered so important to public health that a host of laws and regulations have been promulgated and are in force to protect consumers from food-borne health hazards (see Chapter 9). Thousands of regulatory professionals (especially medical and veterinary) working for federal, state, and local governments monitor and in some cases directly supervise food industry operations from farm production through food processing and preservation and distribution. Nonetheless, it is the food scientist who most frequently must establish and operate quality-control programs aimed at assuring the safety of processed foods as well as compliance with the government regulations. Thus he or she needs to understand the guiding principles as well as the practical means that should be employed to guarantee food safety. Some of the more important principles involved will be examined next.

Raw Materials Selection and Handling

Quite naturally, the place to start on an effective quality-control program is to attempt to eliminate or at least to minimize any health hazards in the food before the raw materials (e.g., food crops, livestock and poultry, milk, and eggs) and other ingredients reach the processing plant. In the case of raw materials from plant sources, cooperative planning and action for this purpose are required among a number of people—farmers, agronomists and other crop specialists, pest control operators, regulatory authorities, others associated with the growing of the crops, and, of course, the food scientist. In the case of meat, poultry, eggs, and milk, cooperation is needed among farmers, veterinarians, regulatory officials, feed company personnel, drug and pharmaceutical house representatives, others associated with the production of livestock, poultry, eggs, and milk—and again, the food scientist. In the case of fish and shellfish, such cooperation involves fishermen, regulatory officials, others concerned with the harvesting and handling of the fish—and the food scientist.

Generally speaking, the primary role of the food scientist in these operations is to monitor those preprocessing activities that affect the safety of the raw materials, collect and analyze samples for infective and toxigenic agents and toxic compounds, and help to establish standards for accepting or rejecting raw materials and ingredients for processing.

Processing and Preservation Practices

There are numerous opportunities during processing and preservation operations to protect and even to enhance the safety of food. A few of the more important steps will be mentioned here.

Sanitation. In and around the processing plant, sanitation is a key element in controlling the safety of processed foods. At the very beginning of operations the raw materials and other ingredients should be sampled and carefully examined for any obvious evidence of contamination. Excessively soiled and damaged produce can usually be detected and rejected. For example, with the supervision of veterinarians, unhealthy animals can be identified and rejected before slaughter and processing.

Another very important phase of an effective sanitation program is maintaining the processing equipment and facilities in a clean and sanitary condition, especially food-contact surfaces. This involves making sure not only that the surfaces are physically clean to the eye and the hand, but also that they are as free of microorganisms as is possible, by using effective sanitizers.

Periodic and "end-of-the-shift" cleanup and sanitizing of equipment and the processing area are essential in such operations.

Pest control in and around the processing plant is also a key element in maintaining sanitation, especially control of insects, rodents, and birds. This generally requires close cooperation between the food scientist and the commercial pest-control operators who usually carry out pest-control programs in food processing plants. Not only is it important to control the pests, which can contaminate the food with hazardous material, but it is equally important to avoid contaminating the food with the chemicals used for pest control. Storage areas for ingredients and supplies are quite vulnerable to infestation by pests and should be carefully monitored and treated. In some cases it is necessary to construct rodent- and bird-proof storage areas in order to provide adequate protection against invasion and contamination by these pests.

Proper management of the liquid and solid wastes from the processing plant is yet another essential element in sanitation. If not promptly removed from the plant area and properly processed, wastes can readily become a serious harborage for insects, rodents, and even microorganisms, which can then invade the plant and contaminate the food. Close cooperation between waste-management personnel, public health officials, and the food scientist is required here.

An extremely important sanitation measure relates to the food handlers working in the processing plant. These people, if carriers of disease, have been demonstrated to be an important source for the contamination of food with pathogens. This has been well documented for staphylococcal and ba-cillus intoxications and for *Salmonella, Cl. perfringins,* and certain viral in-fections. Control of this source of contamination calls for a rigorous program of medical examination and supervision of the workers handling the food raw materials and ingredients as well as other unprocessed materials. In certain other operations in the processing plant, especially for foods composed of animal products that are to be sold in chilled, frozen, or dehydrated form, the health of the workers is most important. However, it is not so important for thermally processed foods because of the high temperatures employed in their preservation, which destroy pathogens.

Thermal Processing. Heat, besides accomplishing other desired effects (e.g., cooking) is a very effective means of improving the safety of many foods. Foremost among these is the destruction of microorganisms causing infections and intoxications and the inactivation of certain nutritional inhibitors. The principles for the thermal destruction of microorganisms in food have been well established by research carried out over the past six or seven decades. They will receive careful consideration here.

Lethality—that is, the killing power of a heat process—depends on four key factors: 1) the type and number of microorganisms present; 2) their phys-iological state, especially age and whether in vegetative or spore state; 3) the properties of the food, especially its chemical composition and pH; and 4) the time of exposure of the microbes to a lethal temperature. More simply stated, this means that (a) the more heat-resistant the organisms are and the greater their numbers, the more severe the heat treatment must be to kill them all; (b) the older they are and the more organisms that are in spore form, the greater the heat treatment must be; (c) the chemical composition and pH of the food may have either a positive or negative effect on the killing power of a given treatment; and (d) the longer the time of exposure to a lethal temperature, the greater will be its killing effect. Of all the properties of the food, pH is perhaps most important. Microorganisms in foods with a pH of below 4.5 (so-called acid foods) are rather easily killed, whereas those in foods with a pH above 4.5 (so-called nonacid foods) are much more difficult to destroy by heat.

Thermal death time (TDT) is defined as the time required to kill microorganisms at a given lethal temperature. Research over the years has provided us with the TDTs for most of the organisms that present health hazards in food. It has been found that the logarithm of the number of survivors is linear with respect to the time the organisms have been held at a lethal temperature. Figure 26 shows this relationship for a hypothetical situation. A family of TDT plots can be obtained using experimental data collected from thermal-

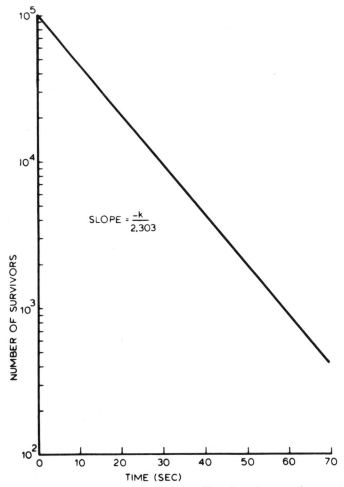

$$\text{SLOPE} = \frac{-k}{2.303}$$

Fig. 26. Destruction of microorganisms by heat. Hypothetical survivor/heating time plot.

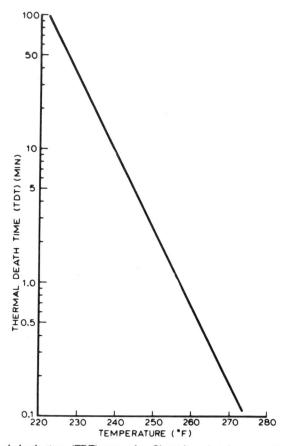

Fig. 27. Thermal death time (TDT) curve for *Clostridium botulinum* in phosphate buffer. (Adapted from Esty and Meyer, 1922. Copyright © by the University of Chicago Press.)

processing studies carried out in the laboratory using special equipment and techniques (Karel *et al.*, 1975). From these plots one can prepare a time/temperature curve for the destruction of the organism. Figure 27 shows the TDTs for *Cl. botulinum* in a buffer solution exposed to a number of lethal temperatures. Similar curves have been published for most other infective and toxigenic microorganisms that occur in foods. Commercial methods used for the destruction of microbes in food will be discussed in Chapter 7 and 8.

Thermal treatment can also be used to destroy certain nutritional inhibitors. In America soybeans are one of the few food raw materials that are commercially heat-treated specifically for this purpose. Raw materials such as lima beans, fish, and eggs, and food products derived from them, usually receive sufficient heat treatment during cooking to render them free, or nearly so, of heat-sensitive nutritional inhibitors.

Refrigeration. A potent means for controlling health hazards, specifically those caused by pathogenic microorganisms and also parasites, is refrigeration. Like microbes, these latter agents have minimum temperatures for growth and in this case, all are within the range of temperatures usually employed in the processing plant, cold-storage warehouse, refrigerated rail cars or trucks, and refrigerated display cases in the retail market. Many of these organisms not only cannot grow at these temperatures but gradually die off.

Freezing temperatures have a further damaging effect on organisms. Many pathogens are killed outright and the remainder die off gradually with time. However, it is important to be aware that not all of the microbes will die at freezing temperatures. If a frozen food is thawed and warmed to a growing temperature, the remaining viable organisms quickly begin to multiply and can soon become hazardous again. Then, too, bacterial spores are very resistant to freezing and they germinate and grow as soon as the requisite temperature is reached.

Chilling results in human pathogens growing poorly as the temperature drops below the optimum for the species—for most types 30°C (86°F) or higher. Below 10°C (50°F), they grow more slowly, and their growth is usually totally arrested below 3°C (37.4°F). Toxigenic bacteria show a similar pattern and so also will not grow and produce toxin at temperatures below 3°C (37.4°F). Certain molds have been shown to produce toxins at temperatures as low as 4°C (39.2°F) and may do so at even lower temperatures. More research is needed on the toxigenic molds to determine exactly the minimum temperature at which toxins will be produced in food.

Chilling has little or no effect on the preformed toxins of bacteria or molds. So, once formed, these toxins will maintain their potency in refrigerated foods for a very long time.

It has been shown that the spores of *Cl. perfringins* and *Cl. botulinum* are highly resistant to freezing, even though their vegetative cells may have succumbed. After thawing and on reaching the requisite temperature, the spores will germinate, grow, and eventually produce toxin.

Viruses and molds are also generally very resistant to freezing as are the preformed toxins produced by molds.

Dehydration (Water-Activity Reduction). All microorganisms require moisture for growth, spore germination, and other essential activities. However, it is not the percentage moisture content per se of food that controls their activities, but rather water activity (A_w). (Water activity is defined as the ratio of the moisture-vapor pressure of the food divided by the vapor pressure of pure water, both at the same temperatures. Values in food vary from close to 1.0 down to a low 0.2 for the very dry foods.) Minimum A_w for microbial activity varies with the type of organisms, the chemical composition of the food, chemical preservatives, pH, oxygen level, and other as yet unknown factors (Troller and Christian, 1978).

Requirements for microorganisms have been specifically worked out for many situations, but much more work is needed before our knowledge will be as complete as it should be for many practical applications. A few examples will be given here where the data are sufficient to warrant making definite statements about the control of growth and/or toxin formation in food using A_w.

Of the toxigenic organisms, S. aureus appears to require an A_w of 0.86 to 0.97 for toxin production depending on the medium; Cl. perfrigins requires 0.93 to 0.97; Vibrio sp. 0.94 to 0.97 for growth. Of the molds, Aspergillus flavus requires an A_w of 0.85 for toxin elaboration. All of these values are subject to revision as more work is done on the specific effects of food composition, pH, oxygen levels, temperatures, and so on. No doubt each type of food will have an effect on minimal A_w values required for a given pathogen to grow. This field of research is extremely active at the present time and within a few years we should have most of the information we need, especially for foods of intermediate moisture content such as the aged cheeses and cured meats and sausages, where the problem of pathogen control is critical. It is important to note that when these foods are moistened or re-constituted and held, the A_w rises, and because many of the microbes present will still be viable, they will grow and thus make the product a health hazard.

Parasites such as Trichinella sp. are quite sensitive to a low A_w and usually die off rapidly when the product is dehydrated, even partially, but especially so in the presence of any curing agents. Government regulations specify the salt content and drying conditions that are required for cured pork products to eliminate this health hazard (see Chapter 9).

Chemical Preservation. A limited number of chemicals are legally permitted in food to control pathogenic organisms. Salt (sodium chloride), sodium nitrate and sodium nitrite, and sugars are the most common ones in use, and these are usually employed in the curing and smoking of meat and fish. Though the addition of these chemicals lowers A_w, they also appear to have a specific microbial effect of their own. For example, recent research findings have revealed that nitrite salts inhibit the production of toxin by Cl. botulinum. A good deal more research on the use of chemicals to control pathogens in our food is needed and is currently under way. This topic will be discussed again in Chapter 8.

Other Means of Control. There is much still to be learned about efficient methods for the effective control of the food-borne diseases. For example, we know that competition between microorganisms present in food exists, so that the growth of one type may suppress that of another. It may be possible, especially with fermented foods, to take advantage of this phenomenon by using cultures that not only produce the desirable flavor, texture, and so on, of the end product, but also that prevent pathogens from growing. Time will tell. The role of special packaging methods also needs to be more fully explored. Food packages are available that permit control of the internal gaseous environment; thus one can have a completely anaerobic environment inside a food package. Unfortunately, this would aggravate the health hazard in some foods by permitting the growth and toxin production by an obligate anaerobe such as *Cl. botulinum.* On the other hand, under other circumstances an anaerobic environment (e.g., nitrogen-gas packaging) might be used to prevent growth of and toxin formation by *A. flavus,* a strict aerobe. Certain kinds of packaging permit the use of gases to control pathogens. Carbon dioxide is a good example of a gas that inhibits the growth of many microbes, and ethylene oxide is a powerful germicide that can be applied "in-package." After its lethal action has taken place, ethylene oxide decomposes into harmless ethylene glycol. These and other possibilities require exploration and are under active investigation at present. We will be discussing some of these applications in Chapters 7 and 8.

REFERENCES

Esty, J. R. and Meyer, K. F. (1922). *J. Infect. Dis.* **31**, 650.

Karel, M., Fenneman, D. R., and Lund, D. B. (1975). "Physical Principles of Food Preservation." Dekker, New York.

Riemann, H., and Bryan, F. L., eds. (1979). "Food-Borne Infections and Intoxications," 2nd ed. Academic Press, New York.

Tannenbaum, S. R., ed. (1979). "Nutritional and Safety Aspects of Food Processing." Dekker, New York.

Troller, J., and Christian, J. H. B. (1978). "Water Activity and Food." Academic Press, New York.

SELECTED READINGS

Ayres, J. C. (1973). Antibiotic, inhibitory and toxic metabolites elaborated microorganisms in foods. *Acta Aliment. Acad. Sci. Hung.* **2**, 285–302.

Ayres, J. C., Mundt, J. O., and Sandine, W. E. (1980). "Food Microbiology." Freeman, San Francisco, California.

Beuchat, L. B., ed. (1978). "Food and Beverage Mycology." Avi, Westport, Connecticut.

Brennen, J. C. (1978). "Basic of Food Alergy." Thomas, Springfield, Illinois.

Catsimpoolas, N., ed. (1977). "Immunological Aspects of Food." Avi, Westport, Connecticut.

Fair, G. M., Geyer, J. C., and Okum, D. A. (1971). "Elements of Water Supply and Waste-Water Disposal." Wiley, New York.

Goldblatt, L. A., ed. (1969). "Aflatoxin." Academic Press, New York.

Graham, H. D. (1980). "Safety of Foods." Avi, Westport, Connecticut.

Guthrie, R. R. (1980). "Food Sanitation." Avi, Westport, Connecticut.

Jowitt, R., ed. (1980). "Hygienic Design and Operation of Food Plant." Avid, Westport, Connecticut.

Kastuyama, A. M. (1979). "A Guide for Waste Management for the Food Processing Industry." Food Processors Inst., Washington, D.C.

Katsuyama, A. M. (1980). "Principles of Food Processing Sanitation." Food Processors Inst., Washington, D.C.

Liener, I. E., ed. (1974). "Toxic Constituents of Animal Foodstuffs." Academic Press, New York.

Liener, I. E., ed. (1980). "Toxic Constituents of Plant Foodstuffs," 2nd ed. Academic Press, New York.

Troller, J. (1982). "Modern Food Plant Sanitation." Academic Press, New York.

Chapter 5

Evaluation of Sensory Properties of Foods

The senses by which we perceive food are sight, smell, taste, touch, kinestheses, temperature, and pain. These senses are often confused or not noticed. We are usually unaware of the differences among taste, feel, and odor sensations when we bite into food. If the nose is tightly closed, the sensation received from eating a peach is little more than sweetness, a slight sour taste, a slight bitterness, and the cooling feel of the pulp. What we commonly ascribe to odor may be mainly taste or touch. For a general discussion of the senses in relation to foods, see Amerine *et al.* (1965), Raunhardt and Escher (1977), and Carterette and Friedman (1978).

Unfortunately, our understanding of sensory physiology (and of sensory psychology) is still only in the developmental stages. When more fundamental data are available, food scientists will be able to make more perceptive measurements of human sensory response to foods. This becomes of critical importance as we enter, apparently, a period of energy deficiency and new, safe, nonanimal sources of food have to be developed. Food scientists must make sure that new foods are not only nutritious but also appealing to the consumer as regards appearance, color, smell, taste, feel, and flavor.

Sensory Factors

Appearance and Color

One of man's earliest prejudices was against dirt on food. This prejudice is now largely visual, but in the beginning it may have been based on the sense of touch and on the avoidance of unplesant contaminants such as grains of sand. Renner (1944) also points out that the color of dirt is usually dark, and such colors are generally considered unpleasant, possibly because mold growth is often dark, as is fecal matter. Color is particularly important in foods because if the color (or appearance) is unattractive, the consumer may never touch or taste the food.

Color Prejudices. Color prejudices among individuals exist, often with a valid basis. Green fruit is usually unripe, whereas yellow and red are associated with ripeness and desirable flavor. A certain color (or lack of it) is associated not only with the native quality of the food but also with its processed quality. Mold spoilage of meat, lack of adequate mold in blue cheese, browness in canned corn, and an amber color in white table wines all reflect negative color-quality factors. The pigment responsible for the color of fresh meat is the purple heme pigment myoglobin. On exposure to oxygen, this forms oxymyoglobin, which is bright red. The oxidized pigment metmyoglobin is a grayish brown. Use of fluorescent fixtures in the meat department display counter shifts the visible color of the meat toward the red, making it appear less oxidized and thus more appealing.

Proper color is often the most important or the only quality factor the consumer recognizes. With alcoholic beverages, color and appearance are widely accepted as an important measure of quality. Cloudy wines are usually spoiled, and off-color wines may be too young or too old or may have been improperly processed. The color and appearance of fish, fruits, and vegetables are often good indicators of quality, particularly their freshness or maturity. The Boston market traditionally paid more for brown eggs than for white, but here the basis of the color–quality evaluation is not known.

The color of food also may appeal because of aesthetic considerations. The combination of colors in salads, the juxtaposition of colors on the plate, and the decoration on cakes all contribute to our appreciation of foods. We judge the weakness or strength of tea or coffee by the color, and we classify wine or beer by color. We insist on colors in some foods that we condemn in others: green egg white would be repugnant, but blue cheese and green and blue plums are accepted and appreciated.

The U.S. Food and Drug Administration (FDA) considers any added color, natural or artificial, to be a food additive. It must be specifically approved for use and be noted on the label. The labeling required depends on whether the color is natural or artificial. Any detectable quantity that may have been added must be noted. Where a standard of identity for a food has been developed, if that standard does not contain a statement of added color, then color additives may not be used. (See Chapters 4 and 9 for further details on food safety and food laws and regulations.)

The eye records our impressions of the physical world: color, size, shape, gloss, glitter, and so on. However, the eye is not a very good quantitative instrument, though it is very fine qualitatively.

Color Measurement. Color cannot be expressed by a single parameter. To define a color we must determine its intensity, dominant wavelength, and colorimetric purity (Hardy, 1936). Roughly, intensity is the amount of the color, the dominant wavelength is the predominant color (e.g., red, yellow), and colorimetric purity is the relative amount of gray present. Either the light reflected from the surface or that transmitted through the food is analyzed. Computer programs now available make possible fast and accurate calculation of color properties. For some foods, color scales that show proper colors for harvesting are available (Francis, 1977). There are two categories of appearance: color attributes related to wavelength distribution, and geometric attributes related to the spatial distribution of light.

Like the other sense organs, the eye has its physiological properties and limitations. The most important of these are 1) absolute threshold, 2) differential threshold, 3) duality of reception, 4) adaptation, and 5) hue and saturation discrimination. The effective light energy at the threshold is a few hundred-billionths of an erg (unit of energy) or 5–11 quanta (average 7). This means that the eye is so sensitive that if it were much more so, we could "see" the "shot effect" of photon emission, and "steady" light would not be steady.

Various systems of color nomenclature are available, and color maps are used to provide reference points. The Maerz and Paul "color dictionary" contains 7000 color samples under 4000 names. The Munsell system is widely used in the United States. Finally, the tristimulus system of the International Commission on Illumination (ICI) is now commonly accepted by food scientists (see Hardy, 1936). Purity (saturation) and brightness (amount of color) are obtained from an absorption curve hue (dominant wavelength).

Color standards are available for many foods. An example is the beef color device of Kansas State University; unfortunately, this does not correlate well with visual reflectance data. There are also charts for tomato color (TC), citrus redness (CR), citrus yellow (CY), and lima bean white color standard (of the USDA), and plastic color tubes for orange juice.

Other Visual Factors. With some foods, factors other than color are important, such as glitter, gloss, sheen, translucency, and the uniformity of these over the surface. Even very secondary visual aspects of foods are important. Floating particles, fat on soup, creaminess of cheese, foaminess of beer, or the shiny surface of some cake frostings are important negative or positive components of their quality. Appearance factors such as physical form, visual texture, shape, size, irregularity of surface (including presence of fibers or granules) and the gross texture or consistency properties (like fluidity) influence consumer response. The visual presentation of the food on the market (such as product description, packaging, and illumination) or on the consumer's plate (colors, size of serving, arrangement of different foods) also influences response (Hutchings, 1977). These are called the geometric attributes of color.

No work seems to have been done on the importance of the various types of color blindness for food appreciation or lack of appreciation.

Smell

The olfactory sense is the second most important sensory guide to food appreciation. Olfaction supplements taste to facilitate ingestion of nutritious foods and to protect against those that are poisonous (Cain, 1978b). The smell of food encourages consumption and activates digestive processes in the mouth and stomach. Unpleasant odors cause rejection; unfortunately, not all poisonous foods have undesirable odors, and some foods with unpleasant odors may be nutritious. (For a perceptive history of odors and health, see Cain, 1978a.) Unpleasant odors often arise from poor sanitary conditions. For the character-impact and contributory odors of some foods see Tables 14 and 15.

TABLE 14
Some Foods Whose Aroma Resides Largely in One Compound[a]

Food	Character-impact compound
Banana	Isopentyl acetate
Grape, Concord	Methyl anthranilate
Grapefruit	Nootkatone
Lemon	Citral
Pear	trans-2,cis-4-Decadienoates
Cucumber	trans-2,cis-6-Nonadienal
Green pepper	2-Isobutyl-3-methoxypyrazine
Potato, raw	2-Isopropyl-3-methoxypyrazine
Mushroom, boiled	1-Octen-3-one
Shiitake	Lenthionine
Beetroot	Geosmin
Garlic	Di-2-propanyl disulfide
Watercress	2-Phenylethyl isothiocyanate
Almonds	Benzaldehyde
Cheese, Blue	2-Heptanone and 2-Nonanone

[a]From Birch et al. (1977a).

TABLE 15
Some Foods Whose Aroma Is Due to a Mixture of
a Small Number of Compounds[a]

Food	Character-impact compound	Contributory odor compounds
Apple, Delicious	Ethyl 2-methylbutyrate	Hexanal
		trans-2-Hexenal
Bilberries		Ethyl 2-methylbutyrate
		Ethyl 3-methylbutyrate
		trans-2-Hexenal
Raspberries	1-p-Hydroxyphenyl-3-butanone	cis-3-Hexen-1-ol
		Damascenone
		α- and β-Ionone
Tangerines		Methyl N-methylanthranilate
		Thymol
Tomatoes		Hexanal
		trans-2-Hexenal
		cis-3-Hexenal
		cis-3-Hexen-1-ol
		2-Isobutylthiazole
		Some high-boilers
Celery	3-Isobutylidene-3a, 4-dihydrophthalide 3-Isovalidene-3a, 4-dihydrophthalide	cis-3-Hexen-1-yl pyruvate 2,3-Butadione
Onion, raw	Thiopropanal S-oxide (lachrymator)	Thiosulfinates Thiosulfonates
Onion, boiled	Propyl and 1-propenyl disulfides	
Cabbage, boiled	Dimethyl disulfide	2-Propenyl isothiocyanate Dimethyl trisulfide
Potato, boiled	2-Ethyl-3-methoxypyrazine	Methional
Butter	2,3-Butanedione	Ethanal Dimethyl sulfide

[a]From Birch et al. (1977a).

Smell is not only a part of the esthetic pleasure in foods; it also seems to serve in special cases as a protective mechanism. Spoiled foods often have typical and easily recognizable odors that cause us to reject them. This is particularly true of meats.

The senses are not, however, very good watchdogs of health. Putrified meats can be and have been eaten. If properly cooked, they usually do not harm health. Until the nineteenth century, butter was usually rancid when served. Spoiled vegetables have been eaten, but in some cases they have caused poisoning. Poisonous mushrooms often have as delectable an odor as the safe ones. (See Chapter 4 for further aspects of this subject.)

Odor Quality. The particular quality associated with an odor has intrigued organic chemists and psychologists. None of the numerous theories proposed has survived unscathed: vibrational, chemical, chemical–vibrational, electrochemical, enzyme, informational, physical, and physicochemical (Cain, 1978a). Inorganic substances, with such exceptions as hydrogen sulfide, are not potent odorants. Substances of high (>300) or very low (<15) molecular weight generally have very little odor. But just what gives rise to the particular stimulus has escaped us. Substances with similar physicochemical properties have quite dissimilar odors, whereas others with different physicochemical properties have similar odors. (See Beets, 1973, 1978, and Ohloff, 1972, for specific data.) Many classifications of odor quality have been made (Cain, 1978a), but none seems to explain all the aspects of odor quality. Pleasantness seems to be the primary parameter. Multidimensional scaling techniques may give us a psychological arrangement of odors.

Odor Potency. The human nose is very sensitive to odor (one part per trillion or less for some odors). In terms of parts per million volume, the recognition thresholds of some common compounds found in food are 0.00047 for hydrogen sulfide (rotten egg), 0.00021 for trimethyl amine (fishy), 0.47 for sulfur dioxide, and 46.8 for ammonia.

The natural aroma of fresh fruits and vegetables or of certain processed foods is one of their most attractive features. In addition, we often add herbs and spices to produce special desirable smells. For example, we add vanilla extract to ice cream and herbs to wine for producing vermouth.

However, standards for odor quality in food are not universal. Decomposed fish sauces are consumed and enjoyed by millions of people in Southeast Asia. "Ripe" cheese is not appreciated by many people. Appreciation of the smell (and taste) of curry must certainly be cultivated.

All odorous materials are volatile, but the volatility (or vapor pressure) of a compound is not proportional to its odor. Musk (used in perfumes) has a low volatility but is one of the most powerful odorants. On the other hand, water has a relatively high vapor pressure but is odorless.

Odors are generally emitted by organic compounds, but the relation of composition to odor is extremely variable. Compounds of very different chemical composition may have quite similar odors, whereas compounds of similar composition may have different odors. In concentrated solutions many compounds have an unpleasant or repugnant odor, but in dilute solutions the odor may be very pleasant. For example, hydrogen sulfide is repugnant at concentrations of more than about 5 to 10 parts per billion, but trace amounts (less than one part per billion) are normal in beer and to a certain extent contribute to the "beer" aroma.

Olfactory Region. The olfactory locus has been established in a sharply localized region in the upper part of the nose (Cain, 1978a). Electrophysiological readings from the olfactory nerves were first recorded about 50 years ago, but single nerves for specific odor qualities have not been identified. Unfortunately, most olfactory cells respond to as many as five odors.

Beets (1978) has noted that the olfactory system detects structural details of odorant molecules, translates the informational content into usable types of information, processes the latter, and delivers "the resulting pattern to the higher centers [of the brain] for comparison with memory contents, for interpretation, and for conversion into a terminal product such as an odor sensation, a verbal expression, or a behavioral effect."

The olfactory process starts with odorant molecules in the layer of mucus covering the membranes of the receptor cells, of which there are millions. Each cell is linked to its own axon. The axons enter the olfactory bulb, which acts as a kind of switchboard, eliminating nonessential information and "noise" before passing the signals on to the brain.

Olfaction is the most elusive and mysterious of the senses. The ability to discriminate among hundreds of different odor qualities and to sense a single stimulus over several log units of concentration is truly astounding.

The air currents produced in normal breathing do not reach the olfactory receptors. Sniffing or irregular breathing will draw air into the region of the superior turbinate so it reaches the olfactory region. The olfactory sense must be carefully distinguished from taste, the receptors of which are on the tongue. This can be easily demonstrated if the olfactory region in the nose is carefully blocked off by pinching the nose tightly.

Measuring odors is difficult because of the inherent problems of accurately measuring the very small molecular concentrations and the variable volumes reaching the nose. Various methods have been devised for this purpose, none of which is entirely satisfactory.

Thresholds. The recognition thresholds for some common chemicals, some associated with food, are given on p. 115. Stahl (1973) has published a list of odor and taste thresholds. Studies vary widely in threshold-concentration findings because of differing methods of stimulus presentation, differences in the solvents used to dilute the odor, size and wetness of the nasal passages, individual sensitivity, differences in method of testing, environment, and other factors (Cain, 1978b). True olfactory sensations must be distinguished from those that excite the trigeminal nerve. Most of the latter cause pungent, tickling, stinging, burning, cool, warm, or painful responses. At high concentrations most, if not all, odors excite the trigeminal nerve, and Cain (1978b) believes that this nerve contributes to the overall odor magnitude, even when the compound concentrations are low.

Adaptation—the decline in sensitivity with continuous odor stimulation—results in a rapid increase in thresholds after breathing a particular odor. The rate of adaptation is a function of the stimulus intensity and is not complete for some time. Cain (1978b) reported that the perceived magnitude of some odors decays about 2.5% per second for weak stimuli and faster for strong stimuli. He believes that olfactory sensations adapt more rapidly than trigeminal sensations. Although complete adaptation (no response) normally does not occur, habituation to odors, even unpleasant ones, does.

With some natural odors, the olfactory sensation appears to be complex, and some constituents fatigue the olfactory mechanism faster than others. Thus as adaptation occurs, the nature of the odor may change. In a few cases, fatigue for one odor may even raise the thresholds of others. For example, adaptation to camphor raises the thresholds for cloves and ether. Whether by mixing two odors a complete neutralization of odor can be obtained is still controversial, but when two odors are mixed, the smell is less than the sum of the unmixed components. This is called compensation. Some masking (neutralization) does occur and is used commercially in deodorizing systems. A few cases of synergism (increase in sensation) are known.

Control of the olfactory environment is an obsession in the United States and other countries, from halitosis and perspiration to slaughterhouses and other food processing plants. Yet odors act as warning agents, are an integral part of our food, and may act as sex attractants or repellents. Clean-air laws have forced most food plants to reduce or partly neutralize malodorous effluents.

Taste

Taste and smell must not be confused: taste refers only to sweetness, sourness, saltiness, and bitterness perceived in the mouth (almost exclusively on the tongue). (See Table 16 for the historical development of the four-modality concept of taste.) All four taste modalities are important in some foods, and sometimes several are found in the same food. All taste stimuli are soluble in water.

The taste sense of the infant seems to be well developed at birth, as can be deduced from facial gestures, tongue movements, modification of sucking patterns, physiological response, and intake measurements (Weiffenbach et al., 1980).

The degree of the taste sensation depends on solubility, ionization (in the case of acids and salts), and temperature. Most of the taste buds (organs of taste) are located in the papillae of the tongue. Children also have taste receptors on the insides of their cheeks, but most of these later disappear or become nonfunctional. People who have lost their tongues are able to taste, apparently because of those taste buds remaining in the cheek.

There is some spatial arrangement of the taste buds on the tongue, so that, for example, one region is somewhat more sensitive to bitterness (the back

TABLE 16
Selected Historical Lists of Taste Quality Names[a]

Aristotle (384–322 B.C.)	Avicenna (980–1037)	Fernel (1581)	Haller (1786)	Rudolphi (1823)	Horn (1825)
Sweet	Sweet	Sweet	Sweet	Unlimited	Sweet
Bitter	Bitter	Bitter	Bitter		Bitter
Sour	Sour	Sour	Sour		Sour
Salty	Salty	Salty	Salty		Salty
Astringent	Insipid	Astringent	Rough		Alkaline
Pungent		Pungent	Urinous		
Harsh		Harsh	Spiritous		
		Fatty	Aromatic		
			Acrid		
		Insipid (tasteless)	Putrid		
			Insipid		

Valentin (1847)	Wundt (1880)	Öhrwall (1891)	Kiesow (1896)	Hahn (1948)	Zotterman (1956)
Sweet	Sweet	Sweet	Sweet	Sweet	Sweet
Bitter	Bitter	Bitter	Bitter	Bitter	Bitter
	Sour	Sour	Sour	Sour	Sour
	Salty	Salty	Salty	Salty	Salty
	Alkaline	Insipid (referred to the flat taste of water)		Alkaline	Water (in some species)
	Metallic				

[a]From Bartoshuk (1975).

of the tongue), another to sweetness (the tip of the tongue) (see Fig. 28) for distribution of taste buds).

Sweetness. Sweetness is particularly important in soft drinks, fruits, and fruit juices, in honey, and in many baked products. Because sugar as such was not widely available to primitive humans, it is only as a cultivated taste that it is important to the quality of many foods. Sweetness is elicited from many types of compounds: salts (lead acetate), alcohols (glycerol), sugars, complex aromatics (saccharin), and organometallic compounds known as cyclamates. Natural sweet-tasting compounds are usually but not always nonionic.

Several sweetening agents besides sugars have been used in food. However, excessive and continuous use of sugar substitutes is not free of danger, especially among people who may develop toxicity (at least to the teeth)

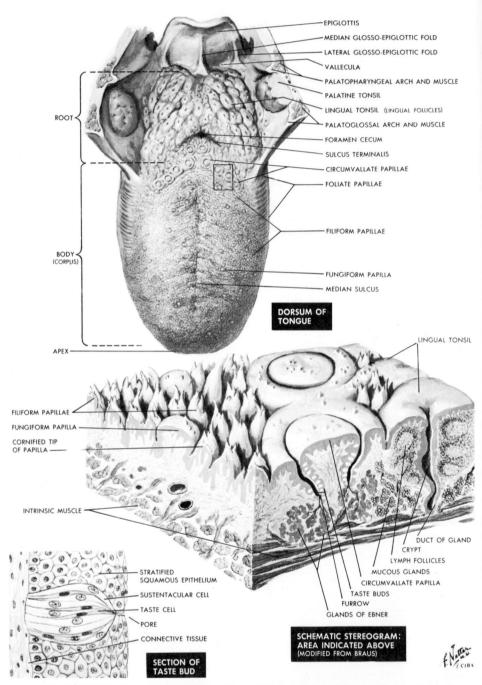

Fig. 28. Distribution of circumvallate, filiform, foliate, and fungiform papillae on the human tongue. (From Netter, 1959. Copyright © 1959, CIBA Pharmaceutical Company, Division of CIBA-GEIGY Corporation. Reprinted with permission from *The CIBA Collection of Medical Illustrations,* illustrated by Frank H. Netter, M.D. All rights reserved.)

caused by their excessive consumption. Because cyclamates are banned in the United States and saccharin is under a cloud, very active research is under way to find a nontoxic nonsugar sweetener. Protein sweeteners such as monellin and thaumatin have been isolated, and they are 20,000 and 1600 times as sweet as sugar, respectively. They are not yet approved for commercial use in this country.

Sourness. Sourness, or the tart taste of acids, is also important in fruits and fruit juices as well as in a number of fermented products such as pickles, sauerkraut, and wines. The lack of a certain amount of acidity results in a flat and unpalatable taste in many foods. (It is doubtful that the acid taste has much protective value for humans in the selection of food. Even the most acid foods are not strong enough in acidity to be injurious to health.) The sour taste is a reaction to ions—primarily hydrogen ions. We should expect, therefore, that pH (hydrogen ion concentration) would give a complete explanation of the acid taste. However, at equimolar concentrations, acetic acid is more acid in taste than hydrochloric acid, although the pH of the latter is lower. This may be due to interactions of the saliva and the acid compound. Minimizing the effect of the saliva does seem to give values for the acid taste that are more directly related to pH (Pfaffmann, 1951). Weak acids, which taste more acidic than would be expected based solely on pH, may indeed elicit tastes other than the simple sour taste. The importance of pH and of the acid taste to biological stability and in food appreciation can hardly be overestimated (see Chapters 4, 7, and 8).

Bitterness. The bitter taste is appreciated in beer, in certain types of wines, and in many other foods. Many poisons are bitter, but the bitter taste can hardly be depended on as a health safeguard, because some very poisonous substances are not bitter. A number of ions, such as magnesium, ferric, and iodide, are bitter. The most typical bitter tastes are those of the alkaloids: quinine, caffeine, and strychnine are intensely bitter. Unlike saltiness and sourness, bitterness resembles sweetness in that it results from the stereochemistry of stimulus molecules. In fact, if sugar molecules are chemically modified, the resulting derivatives are almost always bitter, sweet, or bittersweet (Birch *et al.*, 1977b).

The threshold curve for most compounds is normal—that is, unimodal— but the curve for the bitter taste of the organic compound phenyl thiocarbamide (PTC) is bimodal (Fig. 29). About one third of Americans get a bitter reaction from this compound, even if it is tasted in very small concentrations. This emphasizes the complicated stimulus mechanism for the bitter taste.

Saltiness. Salt appears to be hedonically negative to the newborn infant, but after about 2 years of age children generally prefer salted to unsalted foods.

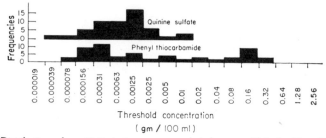

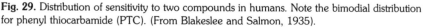

Threshold concentration
(gm / 100 ml)

Fig. 29. Distribution of sensitivity to two compounds in humans. Note the bimodial distribution for phenyl thiocarbamide (PTC). (From Blakeslee and Salmon, 1935).

We like the salt taste on products as diverse as meat and cantaloupes. A somewhat better case for the necessity of a salt taste sense in humans can be made than for sugar, because we have a physiological need for salt. In the case of saltiness, both the anion (Cl^-) and cation (Na^+) have an influence. Potassium and calcium chlorides have a salty taste, but it is different from that of sodium chloride, the typical salty compound. Likewise, sodium fluoride and iodide are salty, but they are somewhat different from sodium chloride. The difference may depend partially on other sensations.

Bartoshuk (1977) has shown that mixing tastes (sweet and sour, for example) does not produce a new taste. The sweet and sour tastes can be recognized in the mixture, albeit with reduced intensity, or one taste may be repressed or intensified.

Recently, application of certain plant extracts (miracle fruit and artichokes) has been found to add a sweet taste to substances that normally taste sour. Another plant, gymnema, selectively abolishes the sweet taste.

Adaptation. Taste is subject to adaptation. Continuous application of taste solutions to individual papillae on the tongue gradually results in insensitivity. Adaptation time for saltiness has been shown to be a function of the concentration of salt in the solution being tasted. We adapt quickly (20 seconds) to low concentrations and less rapidly (2 minutes) to high concentrations. The rate of adaptation for different compounds varies, as does the recovery rate. (Both rates are faster for salt than for glycine.)

Taste sensitivity varies with individuals and with temperature. The situation for PTC (p. 113) has already been mentioned. Salt and quinine thresholds increase with temperature; that of hydrochloric acid remains constant from 62.6° to 107.6° F (17°–42° C); and that for dulcin decreases from 62.6° to 95.0° F (17°–35° C) and rises slightly at 107.6° F (42° C). These facts are important in determining taste thresholds and are of considerable interest to food processors in establishing seasoning combinations and levels. The fact that temperature also affects adaptation suggests that it is peripheral (at the level of the taste buds) rather than central (in the brain). More research on the effects of temperature on tastes is needed.

Thresholds. The average detection thresholds of some common chemicals are shown in Table 17. These concentrations, except for quinine sulfate, are much higher than those for the sense of smell. However, the ability to differentiate two concentrations of a taste is of about the same order as that for smell.

Interactions. The effect of one taste on another is usually a desensitizing effect (i.e., an increase in threshold). Some of these effects may have a chemical origin. For example, salts generally reduce the sourness of acids, whereas certain acids increase saltiness. This is a subject of great interest to food technologists, and much more work should be done on taste interactions. A curious aspect of taste is the interaction of tastes that takes place on the tongue. If an imperceptible concentration of one compound is applied to the tongue and then another contrasting taste is applied, the first substance becomes perceptible. At higher concentrations, the second taste generally reduces sensitivity to the first. Not all people react the same when tasting two compounds together.

Water can taste sweet, sour, bitter, or salty, depending on the nature of the substance that preceded it on the tongue (Bartoshuk, 1975). However, the most common taste quality of water is sweet. One example is the sweet taste of water after eating globe artichokes (not all people react in this way). The taste of water is important when measuring taste thresholds of other compounds, particularly at low concentrations.

When tastes are mixed, each one shows less intensity than when tasted separately. This is particularly true with bitterness, sweetness, and saltiness. The taste intensity of some complex foods might well be unpleasant were it not for this phenomenon.

TABLE 17
Average Detection Thresholds of Some Common Chemicals[a]

Compound	Concentration at threshold	
	Percent	*Molar*
Sucrose	0.7	0.02
Sodium chloride	0.2	0.035
Hydrochloric acid	0.007	0.002
Saccharin	0.0005	0.00002
Quinine sulfate	0.00003	0.0000004

[a]From Pfaffman (1951).

Touch

The sense of touch is often underestimated as a factor in the appreciation of foods. However, consider the importance of the creaminess of chocolate, the graininess of nuts or pears, the crumbliness of cake, and the sogginess of wet bread. Touch seems to have little protective value so far as our recognition of the safety of the food is concerned, but it does help us to reject sharp or bulky foods.

A variety of surface phenomena are important in food appreciation: contact pressure, deep pressure, prick pain, quick pain, warmth, cold, heat, muscular pressure, tendinous strain, appetite, hunger, thirst, nausea, sex, and so on. Not all of these are touch phenomena. Whether itch, tickle, suffocation, vibration, and satiety are independent senses is not known. The body's mechanism of responses is poorly understood, which may account for the difficulty of identifying the different touch senses.

One subdivision of surface sensations is between the surface feels (pressure, pain, and temperature) and the internal sense in the deeper tissues, such as the muscles. Adaptation to highly localized pressure takes longer than adaptation to generalized, diffuse pressure.

Other skin discriminations, many of which are of great importance in quality appreciation of particular foods, are hardness and softness, roughness and smoothness, wetness and dryness, stickiness, and oiliness. Slight disturbances on a sensitive area set up vibratory impressions. Wetness seems to be associated with cold, hardness with an even, cold pressure having a well-defined boundary, softness with an uneven, warm pressure of poorly defined boundary; stickiness with a variable moving pressure, and clamminess with a cold softness with movement, accompanied by unpleasant imagery. With such a complex sensation and so many aspects of texture, it is no wonder that objective physical methods for texture measurement of foods have still not been highly developed. Methods in use include testing resistance to deformation, simulating the action of the jaws, and measuring physical properties. These tests are important in quantifying the crispness and brittleness of crackers, the toughness of meat, and the ripeness of fruits. The common devices used are the Warner-Bratzler and L.E.E. Kramer shear presses, the Voledkevich bit tendrometer, the General Food Texturometer, the U.S. Army, CSIRO, and Armour penetrometers, and various modifications of these. It is probable that enough data are available to develop tentative standards for texture of many foods (Brennan and Jowitt, 1977; National Academy of Sciences, 1976).

Pain

The pain threshold has been determined, and the distribution of pain sensitivity over the body is well known (Dallenbach, 1939). However, the lower half of the uvula and the region of the inner cheek opposite the second lower molar (Kiesom's area) are completely insensitive to pain. Adaptation to pain takes place just as for the other senses. The attraction of pepper and various capsicums as spices is apparently due to pain reactions; they also have a negative aspect at extreme levels. Individuals vary greatly in their sensitivity to and appreciation of pain in foods.

The "chemical" sense is the response of free nerve endings to chemicals, especially to volatile compounds such as chlorine, hydrochloric acid, sulfur dioxide, and menthol. This so-called sense is another aspect of the pain sensation. These are negative quality factors, except that we usually appreciate the cooling sensation of menthol. The chemical sense may have some protective function in food selection but, considering the hotness of curry, chile con carne, and Tabasco sauce and the number of people who appreciate them, this is doubtful. Spices and some alcoholic beverages may stimulate mild pain and apparently, after adaptations, this response seems to be appreciated for itself.

Kinestheses

The receptors stimulated mainly by the activities of the body are called proprioceptors. These are located in the subcutaneous tissues, in the walls of deep-lying blood vessels, in muscles and tendons, in the coverings of bones, and at the articulations of bones. The stimulation is called "deep" sensitivity or the kinesthetic (literally, "feeling of motion") sense. The muscles of the jaw have this particular sense. Our appreciation of the crispness of lettuce, the brittleness of candy, the crunchiness of certain breakfast cereals, or the crackliness of nuts seems to depend on this sense. Reitz (1961) has emphasized the importance of the kinesthetic sense in food appreciation. As with others, this sense has an absolute threshold. It is also remarkably discriminatory to a small motion displacement, indicating low differential thresholds.

Temperature

Temperature is often related to quality for its own sake, such as in ice cream, hot coffee, cold soft drinks, and beer. Temperature also modifies odors and tastes, which may affect overall appreciation. Extremes of temperature even approaching pain are sometimes appreciated. There are two distinct temperature effects: one is related to the physical or volatility changes brought on by the lowering or raising of temperature and the other to the direct warm or cold sensation of the food. Cold and warm sensitivities are not distributed equally over the body. There are generally more cold spots per square centimeter than warm ones, but the warm-sensitive end organs seem to lie deeper than the cold ones. However, there is a certain ticklishness involved in the responses that makes such mapping somewhat unreliable. Thresholds for warmth and cold can be demonstrated, and complete adaptation occurs at high temperatures in a matter of seconds. Stimulation with heat, however, sensitizes the skin to cold, and vice versa.

Sound

Sound is not of great importance to food scientists but should not be dismissed as unimportant. Reitz (1961) believes, for example, that the sound of the crunching of nuts is an important sensory impression. His unproven theory is that the sound of the crushing of the nuts distracts our attention from their taste and smell and hence increases taste and odor thresholds. The crunchy sound of eating potato chips has been considered of sufficient importance to quality by one company to be used in advertising. Presumably, crunchy potato chips are also crisper.

Flavor

The combined sensory impression of odor, taste, temperature, pain, and touch is generally called flavor. It is, therefore, a very complex sensation. Many food products are specially flavored: carbonated drinks, curries, and vermouths, for example. Usually a flavor base (root beer, cola, lemon) is prepared for such products. Such flavorings must, of course, fit the physical system in which they are to be used (solubility and dispersibility), must be reasonably resistant to changes due to exposure to heat or light, must impart a desirable physical appearance, must be free of spoilage organisms and resistant to spoilage, and, whether natural or synthetic, must have FDA approval (see Chapter 9). One aspect of flavor that is often neglected is appropriateness: the flavor we seek in one food may be loathed in another.

Flavor, visual, tactile, and other sensory impressions affect consumer acceptance. However, what consumers will accept, especially if they are hungry, may not be what they actually prefer to eat. Also, when we realize that cultural prejudice and intellectual activity also influence food preference and food habits, we see how difficult is the problem of defining food quality. Yet this is a fundamental problem for the food scientist. There is no reason to prepare new types of foods or to modify old ones if the consumer rejects them.

If food preferences were the sole determinant of food intake, college students would have a low dietary intake of vitamin A, according to Einstein and Hornstein (1970). In fact, the best sources of vitamin A in their study were among the most disliked foods. People who do not like orange juice or milk and milk products can develop deficiencies of vitamin C and calcium.

If foods are not liked, they will not be eaten, and the nutritional value of any food is zero unless it is consumed.

The professional field devoted to the establishment of criteria for food quality has developed into a large one, chiefly since World War II. New techniques for measurement of food quality are always being developed. This field combines physiology, psychology, and biochemistry (Amerine *et al.*, 1965).

Physiological Factors

Hunger and Appetite

Hunger appears to be primarily a physiological phenomenon. Over 100 years ago, Weber suspected that contraction of the muscle fibers of the empty stomach was responsible for hunger. In 1912, the Cannon–Washburn technique demonstrated experimentally the coincidence of hunger pains with powerful stomach contractions.

However, hunger is more than muscle contraction, for when the stomach is entirely removed and the esophagus joined directly to the intestines, hunger is still perceived. Also, not all stomach contractions produce hunger pains.

The desire for food is called appetite, and this sensation has psychological aspects related to but not necessarily identical to hunger. Young (1957) claimed it is misleading to assume that food intake measures a single variable called "appetite." Palatability and affective arousal following ingestion (or being deprived of either of these), existing habits and attitudes, and the chemical state of the organism (as determined by physiological constitution and dietary history), as well as other complex physiological responses, all influence intake.

Appetite and food intake are controlled by integrative mechanisms (Stellar, 1976). This includes a physiological pathway responsible for regulation and motor mechanisms responsible for motivated and hedonic behaviors. Unfortunately, the physiological signals and neural receptors used for initiation and termination of feeding are poorly understood. The signals come from the eye, mouth, pharynx, and esophagus as well as from the stomach and duodenum.

Dichter (1964) believes that appetite is enhanced by an atmosphere of love, trust, and security. He suggests that it may also arise from a desire to communicate and may be influenced by appearance, color, and shape of foods and by the presence or absence of connections with social aspirations or rituals; perhaps it is also a compensatory gratification. Certainly appetite is important in initiating and maintaining food intake. Many physiological and biochemical factors influence appetite: glucose, lipids, and amino acid pattern in the blood; energy expenditure; specific dynamic action; flavor, taste, and texture of foods; and adequacy of diet. Of course, emotional and psychological stress may also intervene, as will the stress of infection, allergies, and surgery.

The opposite of appetite is repletion or satiety—the gastric and visceral warnings that we have eaten sufficiently. Entrance of food into the duodenum appears to signal satiety (see Smith and Gibbs, 1976, for further information).

Hunger pains are satisfied by a few mouthfuls of food. Nonnutritive and even indigestible materials will help allay hunger pains. Smoking, tightening the belt, swallowing hard, or emotional stress lessen hunger pains. Blood from a starved dog transfused to a satiated dog will produce vigorous gastric contractions in the latter, and vice versa. Because this is not wholly due to blood sugar levels, some unknown blood constituents must have an effect on hunger. There seems to be some specific hunger for vitamins, minerals, proteins, and certain amino acids. With deprivation of these their absolute threshold appears to be lower—that is, we are more sensitive to them.

Obesity is a major dietary problem. Why do people eat too much? Are they hyperresponsive to the palatability of food? Are they reacting to stress? Do they have more numerous food preferences? Could their overeating be attributable to conditions during early training? In some cases obesity is influenced by individual and psychosocial factors that lead to a dysfunction of appetite regulation. Highly tasty food also leads to overeating.

Thirst

Thirst is a reflection of dryness of the mouth and throat. This can be demonstrated experimentally. For example, subcutaneous injection of atropine reduces the output of the salivary glands and induces thirst sensations, and flushing the mouth with water reduces the symptoms. The effect of certain foods on thirst requires further study. Salty and hot foods seem to produce thirst, and some alcoholic beverages may induce it. Alcohol's dehydrating effect may be the cause.

As with hunger, there is a pattern of sensation that produces thirst. Thirst is partially alleviated by the first drink of water. There is also complex motivating force, much like appetite, that informs one as to how much water to drink. All animals maintain a delicate balance of water by this mechanism, possibly through sensing the water balance in the blood. The strategic position of the salivary glands in the mouth appears to give them some control over thirst.

Other Influences on Food Intake

Siegel (1957) showed that when many food items were eaten repeatedly, their palatability rating decreased. The amount of food left uneaten was correlated significantly with palatability rating. The effect dissipated slowly, if at all, within 3–6 months. When diets were monotonous, a greater amount was ingested of items that were rated high initially.

Monotony is a major dietary problem for explorers and sometimes for the very poor. However, if the diet is nutritious and cheap, a society may come to prefer it (e.g., rice in southern China, porridge in Scotland, and the potato in Ireland). In some cases, a small amount of another food may break the monotony, such as salt, sugar, spices, butter, milk, meat, or fish sauces.

The newborn baby can differentiate between pleasant and unpleasant tastes; taste sensitivity to sweet and sour apparently declines in the aged. Nutritional status may also be important. Zinc deficiency is associated with reduced taste sensitivity. Poeple with diabetes mellitus show increased sensitivity to glucose. Lack of insulin plays a role in initiating food intake. Adipose (fat) tissue signals the amount of stored energy and may thus influence food intake.

Sensory Testing

The production of finished foods by the farmer or food processor implies that the consumer will accept these products and pay the requisite price— that is, that the product has a certain quality. It is easier to recognize quality than to define it. Quality is obviously some sort of mental summation of the physical and chemical properties of the food. Many sensory factors are involved, but the relation of each to palatability is not known. So chemical and physical tests, although they give much useful information and frequently can be correlated with quality, must be supplemented with human sensory tests. (See Amerine *et al.*, 1965, and Kramer and Twigg, 1970, for a full discussion.)

Test Procedures

To determine differences between foods, a variety of sensory testing procedures have been developed. These are used to select sensitive panels, to ensure uniform quality, or to detect the difference in food quality between processes, raw materials, or storage conditions. Paired, duo–trio, and triangular tests are widely used for difference testing. In the paired test two samples are presented, and the difference should be specific, such as more and less sweet. The question asked is, "Which sample is sweeter?" Obviously the taster has a 50% chance of selecting the correct "sweeter" sample by chance. To determine whether the difference between the samples is real and not chance, the test must be repeated several times. Appropriate statistical analyses will then reveal how much confidence we can have in the results.

In the duo–trio test, the difference need not be specific. In these tests a standard is presented first and then again as one of two unknowns. The question asked is, "Which sample is the same as the standard?" The chance of choosing the correct sample is again fifty–fifty and, as before, the test must be repeated several times and statistical measures of significance applied.

In the triangular test, three samples are presented, two of which are the same. The question then asked is, "Which is the odd or different sample?" Obviously the two samples (A and B) can be arranged in various ways in the three containers (AAB, ABA, BAA, BBA, BAB, ABB). In this case, the taster has only a one-in-three chance of selecting the correct "odd" sample by chance. As in the other procedures, repetition is needed to permit statistical analyses of the significance of the results.

Ranking and Scoring. In many cases, more than two samples must be tested. Here ranking and scoring procedures are often used. Ranking is simple, but the samples must all be ranked on the same quality (e.g., taste, color, flavor). Furthermore, ranking does not tell us how much difference there is between samples. The first sample may be very good and all the rest very poor, yet they rank 1, 2, 3, etc.

Scoring likewise presents problems of making certain that all of the tasters are using the same factors for scoring. The results, however, can be statistically evaluated and the significance of differences between samples determined. In other words, scoring can be used to rank samples. Whether the value assigned to each component of the total score is correct is not easy to establish, and the additiveness of the separate scores is questionable.

Because of these problems, hedonic scoring has been used. In this score-card, the samples are rated as to the degree to which they are liked or disliked.

The flavor profile procedure is still used, often with modifications. The profile of a food is decided by a panel decision. The panel is highly trained with regard to a particular food. As a group, it determines the order in which the different flavors and odors appear, and their amplitude (roughly, intensity). The results cannot be easily subjected to statistical analysis, but orginators of the procedure claim good replication of results. This is a descriptive procedure, and the results are valuable because they may reveal clues as to minor aspects of quality that would not be revealed by the other tests.

Magnitude estimation is a subjective measuring procedure. The subject estimates the magnitude of the sensation at one concentration and at another. The ratio of the magnitudes reflects the ratios of their perception. Magnitude estimation is used to rescale other (empirical) scales, to discover and develop perceptual laws for food sensations, and to establish sensory-tolerance curves, for example (see Moskowitz and Chandler, 1977, for further information).

Consumer testing requires large panels, from 100 to 1000 or more panelists, selected at random from among the prospective consumer population. Single-sample and paired tests are often used. In single-sample tests, only one product is presented to the consumer at a time. Hedonic or other methods of scoring may be used to determine the degree of acceptability. In addition, most consumer-testing programs attempt to obtain from the consumer's responses some indication as to why subjects voted as they did.

Psychological Factors

Many psychological errors may arise in the sensory evaluation of foods. Individuals vary as to how they tackle a sensory-evaluation problem. We know very little, for example, about how our personality traits influence our judgments.

People also differ in their expectation of what the sensory quality should or might be. This in turn may subtly influence the perceived sensation(s).

Some panelists are less attentive than others, lack interest, or suffer from boredom. To motivate the panelist to achieve better and consistent results, suitable rewards are useful, such as time off, prizes, monetary payoffs, and publicity.

Errors. Many purely psychological errors may influence a panelist. Among these are the error of habituation, that is, continuing to give the same result although the stimuli in successive samples may have increased or decreased. The error of expectation produces just the opposite effect. The panelist is overanxious to find a difference in successive samples and does so, even though none may be present. This is a type of simulus error. If the panelist knows (or thinks he knows) that a food is from a certain producer, is of a certain age, or has been processed in the preferred manner (or vice versa), he may react as if it were important to the quality of the food. The logical error is another form of the stimulus error. When the panelist knows that two food samples have been processed in the same way, he may rate them equally when in fact they may be quite different. The error of leniency arises when the panelist (consciously or unconsciously) wants to praise the food product of some friend.

When scorecards are used, the error of central tendency may occur, especially if panelists are inexperienced. If a 10-point scorecard is being used, the timid, inexperienced judge sometimes gives all the foods scores within a narrow range, such as 5, 6, 7.

Judgment of food quality may depend on whether a pleasant or unpleasant food preceded the sample. This is called the error of effective contrast. If a poor-quality sample is served just after a very fine one, the poor-quality food will be marked lower than it would have been if a high-quality food had not been served just before.

Finally, there is the time-order error. The first sample served is rated higher than the second, even though it may have been of equal or higher quality. This is not universal; there are reports of bias in favor of the second sample. Use of conditioner samples has been recommended.

All of these so-called errors are very real and can disturb or invalidate the results of a sensory evaluation. As far as possible, trained judges should be used in quality control and in studying processing procedures. Obviously, however, in studying consumer reaction, actual consumers must be used.

Setting Up Sensory Tests

The environment for making the sensory evaluation must be suitable: a well-lit room, neutral (gray) walls, air conditioning, spittoons (with running water), and comfortable chairs are helpful. The sensory-evaluation room should be quiet, and panelists should not be allowed to communicate with each other or with the server. Smoking and use of perfume should be avoided. A smelly chemical laboratory is not a desirable place for sensory evaluation. Nowadays special booths (Fig. 30) are used so that one person can serve five booths at a time. By cutting off the panelist's vision of the taking and serving of samples, one avoids unduly influencing the panelist.

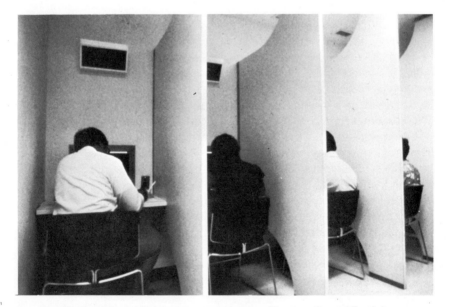

Fig. 30. Sensory panel at work in special booths. (Courtesy General Foods.)

Panel Selection. Selection of panelists requires great care. Not everyone is interested in sensory evaluation. A few have low sensitivity (or cannot smell or taste) or lack familiarity with food. It is wise to eliminate not only the insensitive or bored but also those subject to colds or headaches, and it is important to have panelists who are available daily. Amerine *et al.* (1965) and others recommend screening potential panelists by using the same food and test variables that will be used in the final experiment. Generally the screening test is made sufficiently difficult so that some, but not all, of the potential panelists will fail. A second screening may then be made. However, even with such screening tests, some panelists behave erratically after they have been selected for the panel. Posting the results of each panelist's evaluation may be useful in motivating them to perform better.

Five to ten judges are adequate for the research laboratory and for plant control or new-product development. Of course, in consumer testing, the more who participate the better the test is. A small, motivated, readily available expert panel costs less, uses up less of the test food, and requires less of the panel manager's time.

Coding. The samples have to be coded before serving. Two-digit codes are used, and they should be taken from a table of random numbers. If the raw materials in processing have changed the color so much as to make distinguishing the samples too easy, the panelists may use very dark glasses or the booth may be illuminated with red or purple light.

Number of Samples. The number of food samples presented to the panel at one time can vary from 5 to 50, depending on the objective of the sensory evaluation and the closeness of the quality level of the foods. If the samples vary widely in quality, it may be necessary to have the panel eliminate the worst samples before conducting the sensory evaluation. Some foods are much more difficult to maintain interest and efficiency in than others (olive oil vs. wine, for example).

Correlation of Sensory and Instrumental Data

Because of the relatively high cost of sensory testing, correlation of instrumental and sensory data has long been sought. This might provide clues as to the mechanism of stimulation and perception and could complement panel work in routine quality control or in product and process development work. von Sydow and Åkesson (1977) note that the data may be parametric or nonparametric, so that complicated equations are required for identifying chemical predictors of sensory qualities.

Obviously the samples should be as nearly identical as possible, except for the test variable being studied. In meat studies, only one animal should be used. There should be no picking up of kitchen odors, and the size of samples should be the same, as should their temperature. Person and von Sydow (1974) identified 95 compounds in the head space of canned beef using gas chromatography. They developed computer models that enabled them to predict sensory scores.

There have been numerous studies correlating chemical or physical data with shelf life, aging, and other factors. Factors that make such correlation difficult include the dynamic character and complexity of the reactions, low levels in the product (causing, however, significant off-flavors), and injudicious selection and interpretation of accelerating techniques as they relate to the normal situation in foods (Erickson and Bowers, 1976).

Tsai (1976) wrote that "it is the user's subjective assessment of the quality of food products that finally determines whether they are acceptable . . . despite the considerable research that has been and will continuously be undertaken on the objective methods for quality assessment, their ultimate effectiveness must be judged by their correlation with the subjective assessment." This is the task of consumer tests.

REFERENCES

Amerine, M. A., Pangborn, R. M., and Roessler, E. B. (1965). "Principles of Sensory Evaluation of Food." Academic Press, New York.

Bartoshuk, L. M. (1975). Taste quality. *Dahlem Workshop* 1, 229–241.

Bartoshuk, L. M. (1977). Modification of taste quality. *In* "Sensory Properties of Foods" (G. G. Birch, J. G. Brennan, and K. J. Parker, eds.), pp. 5–25. Appl. Sci., London.

Beets, M. G. J. (1973). Structure-response relationships in chemoreception. *In* "Structure–Activity Relationships" (C. J. Cavallito, ed.), International Encyclopedia of Pharmacology and Therapeutics, Vol. 1, pp. 225–295. Pergamon, Oxford.

Beets, M. G. J. (1978). Odor stimulant structure. *In* "Handbook of Perception. Vol. 6A: Tasting and Smelling" (E. C. Carterette and M. P. Friedman, eds.), pp. 245–255. Academic Press, New York.

Birch, G. G., Brennan, J. G., and Parker, K. J., eds. (1977a). "Sensory Properties of Foods." Appl. Sci., London.

Birch, G. G., Lee, C. K., and Ray, A. (1977b). The chemical basis of bitterness in sugar derivatives. *In* "Sensory Properties of Foods" (G. G. Birch, J. G. Brennan, and K. J. Parker, eds.), pp. 101–110. Appl. Sci., London.

Blakeslee, A. F., and Salmon, T. N. (1935). Genetics of sensory thresholds—individual taste reactions for different substances. *Proc. Natl. Acad. Sci. U.S.A.* **21**, 84–90.

Brennan, J. G., and Jowitt, R. (1977). Some factors affecting the objective study of food texture. *In* "Sensory Properties of Foods" (G. G. Birch, J. G. Brennan, and K. J. Parker, eds.), pp.

227–246. Appl. Sci., London.

Cain, W. S. (1978a). History of research on smell. *In* "Handbook of Perception. Vol. 6A: Tasting and Smelling" (E. C. Carterette and M. P. Friedman, eds.), pp. 197–229. Academic Press, New York.

Cain, W. S. (1978b). The odoriferous environment and the application of olfactory research. *In* "Handbook of Perception. Vol. 6A: Tasting and Smelling" (E. C. Carterette and M. P. Friedman, eds.), pp. 277–304. Academic Press, New York.

Carterette, E. C., and M. P. Friedman, eds. (1978). "Handbook of Perception. Vol. 6A: Tasting and Smelling." Academic Press, New York.

Dallenbach, K. M. (1939). Smell, taste and somesthesis. *In* "Introduction to Psychology" (E. G. Boring, H. S. Langfeld, and H. P. Weld, eds.), pp. 600–626. Wiley, New York.

Dichter, E. (1964). "Handbook of Consumer Motivation: The Psychology of the World of Objects." McGraw-Hill, New York.

Einstein, M. A., and Hornstein, I. (1970). Food preferences of college students and nutritional implications. *J. Food Sci.* **35**, 429–436.

Erickson, D. R., and Bowers, R. H. (1976). Objective determination of fat stability in prepared foods. *In* "Objective Methods for Food Evaluation: Proceedings of a Symposium, November 7–8, 1974" pp. 133–144. Natl. Acad. Sci. Washington, D.C.

Francis, F. J. (1977). Colour and appearance as a dominating sensory property of foods. *In* "Sensory Properties of Foods" (G. G. Birch, J. G. Brennan, and K. J. Parker, eds.), pp. 27–41. Appl. Sci., London.

Hardy, A. C. (1936). "Handbook of Colorimetry." MIT Press, Cambridge, Massachusetts.

Hutchings, J. B. (1977). The importance of visual appearance of foods to the food processor and the consumer. *In* "Sensory Properties of Foods" (G. G. Birch, J. G. Brennan, and K. J. Parker, eds.), pp. 45–56. Appl. Sci., London.

Kramer, A., and Twigg, B. A. (1970). "Quality Control for the Food Industry," 3rd ed. Avi, Westport, Connecticut.

Moskowitz, H. R., and Chandler, J. W. (1977). New uses of magnitude estimation. *In* "Sensory Properties of Foods" (G. G. Birch, J. G. Brennan, and K. J. Parker, eds.), pp. 189–210. Appl. Sci., London.

National Academy of Sciences (1976). "Objective Methods for Food Evaluation: Proceedings of a Symposium, November 7–8, 1974" pp. 7–27, 145–153, 207–213. Natl. Acad. Sci., Washington, D.C.

Netter, F. H. (1959). "The CIBA Collection of Medical Illustrations," Vol. 3, Part 1. CIBA, New York.

Ohloff, G. (1972). Odorous properties of enantiomeric compounds. *In* "Olfaction and Taste." (D. Schneider, ed.), Vol. 4, pp. 156–160. Wiss. Verlagsges., Stuttgart.

Person, T., and von Sydow, E. (1974). The aroma of canned beef: application of regression models relating sensory and chemical data. *J. Food Sci.* **39**, 537–541.

Reitz, C. A. (1961). "A Guide to the Selection, Combination, and Cooking of Foods," Vol. 1. Avi, Westport, Connecticut.

Renner, H. D. (1944). "The Origin of Food Habits." Faber & Faber, London.

Siegel, P. S. (1957). The repetitive element in the diet. *Nutr. Symp. Ser.* **14**, 60–62.

Smith, G. P., and Gibbs, J. (1976). What the gut tells the brain about feeding behavior. *Life Sci. Res. Rep.* **1**, 129–139.

Stahl, W. H., ed. (1973). "Compilation of Odor and Taste Threshold Values." Amer. Soc. Test. Mater., Philadelphia, Pennsylvania.

von Sydow, E. and Åkesson, C., (1977). Correlating instrumental and sensory flavour data. *In* "Sensory Properties of Foods" (G. G. Birch, J. G. Brennan, and K. J. Parker, eds.), pp. 113–127. Appl. Sci., London.

Young, P. T. (1957). Psychologic factors regulating the feeding process. *Nutr. Symp. Ser.* **14**, 52–59.

Chapter 6

Human Nutrition and Food Science and Technology

Historical Background

Food science has been defined as the application of the physical, biological, and behavioral sciences to the processing and marketing of foods. Although the main emphasis in food science is technological, the nutritional aspects should not—indeed must not—be neglected. It must be remembered that food is eaten primarily to satisfy the needs of the body for nutrients. These facts make it clear that food scientists should have a basic understanding of human nutrition if they are to carry out properly the job of converting raw agricultural and fishery products into nutritious as well as acceptable processed foods.

The field of human nutrition, especially the aspects most relevant to food science and technology, covers 1) the history of nutritional knowledge; 2) individual nutrients and their physiological functions; 3) the nutrient content of foods; and 4) the maintenance and even the improvement of nutritive values during processing, preservation, and distribution. Explorations into nutritional knowledge will help in orientation to the field and will also make it possible to learn how nutrition and food science come together and how their integration works for the benefit of man.

129

Early History

Nutrition as a science is of recent origin. Even today our knowledge of human nutrition is meager, as we shall soon see. As one looks back over the thousands of years of recorded history, it soon becomes evident that the development of nutritional knowledge had to await the development of science as a whole, especially in the fields of chemistry, biochemistry, and animal physiology. Before that time, the field was clouded by folklore, superstition, philosophy, religion, and, most important, personal experiences and observations, some wise and some unwise. Nutrition, at least for many laypeople, has not yet rid itself of all these obscuring shrouds! Only since the beginning of the twentieth century has there been any significant emergence of nutrition as a science. (see U S Department of Agriculture, 1959).

The lack of development of the science of nutrition over the years has not been due to any lack of human nutritional problems. In fact, there is ample evidence in the earliest available written records to suggest that humans have suffered nutritional dificiencies from the very beginning of their existence. Not only was there hunger and starvation from the mere lack of food, but also such effects of malnutrition such as night blindness, pellagra, rickets, and scurvy. Drummond (Drummond and Wilbraham, 1958), the noted English nutritionist, has pointed out that night blindness was recognized by medical men in Egypt as early as 1400 B.C. Papyrus documents from this period may be seen today in the Leipzig Museum in Germany describing vitamin A deficiency accurately and correctly prescribing beef liver as a curative! Darby *et al.* (1977) have described other deficiencies found in ancient Egypt.

Evidence for the very early existence of vitamin D deficiency is to be found in the writings of the Greek historian Herodotus (ca. 525 B.C.). He compared the bone structure of the Greeks and Persians of his day and found the latter to be stronger. Perhaps he thought (we do not know this for sure) that this difference was due to the differences in amount of sunshine in the two countries. Much later (about 1700 A.D.) we find numerous references to scurvy, a vitamin C-deficiency disease characterized by severe weakness, loss in weight, generalized hemorrhaging, and bleeding gums. In fact, this deficiency disease very seriously hampered explorations by seagoing ships for hundreds and perhaps even thousands of years. For example, in describing his famous trip to the East Indies in 1497, Vasco da Gama recorded that 60 of a total of 100 of his men died of scurvy during the trip around the Cape of Good Hope. Similarly, Magellan, in his first trip around the world, lost many of his men from scurvy. It was not until about 1750 that the juice of limes and lemons was found by the Scottish physician Lind to prevent scurvy.

By 1650, rickets (a nutritional deficiency disease of the young caused by a lack of vitamin D, calcium, and phosphorus and characterized by soft bones and bowed legs) was so widespread in England as to be considered a prime subject for research in medical schools. Though there was some thought at the time that diet might be important in preventing or curing this condition, it was not until about 200 years later that cod liver oil was recognized as a preventative.

Another nutritional disease of long standing is beriberi. In its terminal phase it is characterized by general malaise, painful rigidity of the body, and finally death. This disease was first observed primarily among eaters of polished rice but was also frequently seen among explorers during long sea voyages. Early Japanese scholars noted that additional meat and fish helped avoid the dreaded effects of this condition. Later on it was recognized to be due to a lack of thiamin, which is present in both whole grains and meat.

From this brief and very incomplete account of the early history of nutritional deficiencies, one can see that humans have experienced a variety of serious disorders since early times and by and large did not learn to cope with them until comparatively recently. To be sure, so long as·people avoided extremes of diet, long sea voyages, and intense industrialization, they seem to have achieved a reasonable nutritional status. However, as population density increased, the pressure on the available food supply also increased to the point where nutritional deficiencies or even starvation appeared. Crop failures due to poor seasons, plant diseases, and insect invasions and similar tragedies with livestock and poultry populations added to human misery and starvation (see Chapter 1).

There seems to be little doubt that industrialization and the accompanying development of cities had a tremendous adverse effect on the quality and quantity of our food supply. Today with a better knowledge of nutritional requirements and the development of a commercial food processing industry, the industrialized nations of the world have been able to overcome this problem and generally to enjoy good nutrition. The developing nations, on the other hand, have as one of their major problems the accomplishment of a similar goal. Only by developing a commercial food processing industry around the world is there any chance of properly feeding all of the peoples of the globe.

Emergence of Nutrition as a Science

The story just related shows that our early knowledge of nutrition was obtained mainly by trial and error. Accomplishments were not as great as they might have been without the interfering influence of superstition, taboos, philosophical dicta, religion, and so on. For example, the great Greek physician Hippocrates (ca. 300 B.C.), often called the father of modern medicine, had many ideas about nutrition and diets, many of which were not sound. His notion that there is one universal nutrient prevailed in some parts of the medical profession until the nineteenth century! Galen, (ca. 130–200 A.D.) another influential physician, wrote extensively about nutrition and health. His ideas, which were largely wrong, were also highly respected. No doubt he adversely influenced the dietary habits of Europeans for centuries.

Nutrition, like agriculture, industry, and medicine, had little chance of significant development until science in general had reached a certain level of sophistication. How could the facts be established and used until the experimental method of science had been developed and accepted? The simple fact is, they could not. Not only had this notion to be accepted, but also science in general had to develop to a point where meaningful nutrition experiments could be made. For example, Sanctorius (a noted scientist of the seventeenth century) studied the changes in his own weight as a result of eating food, of body exertions, and other variables (Fig. 31), but although he asked the right questions about the meaning of the changes in weight he observed, he could not answer them. The sciences of chemistry and physi-

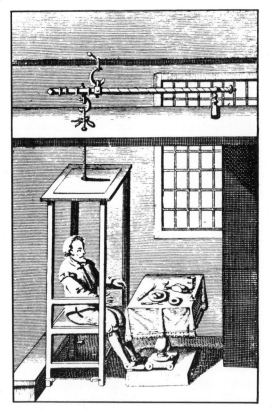

Fig. 31. Sanctorius weighing himself. (From Sebrell and Haggarty, 1967.)

ology simply had not developed to the point of providing meaningful experiments in nutrition until the middle of the nineteenth century, almost 200 years after Sanctorius' time.

Fortunately, the period following the sixteenth century saw a spectacular development of the sciences of chemistry and physics and the beginnings of anatomy and physiology. The Dutch scientist van Leeuwenhoek developed the microscope during this period; the Englishman Harvey demonstrated that blood circulates; and Rutherford discovered nitrogen, Priestley, oxygen, and Black, carbon dioxide. Then the French chemist Lavoisier showed that respiration is the essence of the life process: oxygen is used up by the organism, and carbon dioxide, water, and heat are produced. This brilliant scientist devised means for measuring the changes in oxygen consumption and carbon dioxide production, as well as in heat production, that occur during respiration.

Other significant developments in physiology also occurred during this period. The French scientist de Réaumur observed digestion in birds, noting that the changes involved occurred in the gastrointestinal tract. The Italian scientist Spallanzani performed similar experiments using himself as the experimental subject and discovered that chemical changes take place during digestion, a new fact for his day.

Hippocrates' theory of the universal nutrient was strongly disputed in 1834 by Prout, who put forward the theory that there are three nutrients in food: saccharine (carbohydrate), oily, and albuminous (nitrogenous) compounds. The nitrogenous compounds soon became the subject of many investigations. The word *protein* was coined in 1838 by Mulder, a noted German scientist. Although he was mistaken about the chemical composition of proteins, the name caught on as a designation for the complex nitrogen compounds of living tissue. The amino acid cystine had already been discovered in 1810 by the English scientist Wollaston. Von Liebig, one of the leaders in German chemistry of the time, made many nutritionally important discoveries and is generally credited with being the founder of the sciences of agricultural chemistry and with Pasteur, of biochemistry.

Modern Nutrition

The first use of experimental animals in nutrition research dates back to the last half of the nineteenth century. Lunin, a Russian scientist, very nearly discovered vitamins in milk while attempting to rear mice on a purified diet of protein, carbohydrates, fats, and minerals, in a combination supposedly simulating the composition of milk. Naturally he did not succeed and the mice died. Unfortunately, because he was so engrossed with the idea that certain minerals were causing the mice to die, Lunin overlooked the possibility that another class of nutrients was present in milk. If he had devoted himself to the latter possibility, vitamins might have been discovered much earlier than they were.

The Dutchman Eijkman has been credited with the discovery of vitamins. He found that the polishings from the milling of rice prevented and cured a neuritis condition in chickens caused by feeding polished rice. It is interesting to note that this scientist had been sent to the East Indies about 1890 by the Dutch government to seek a cure for beriberi in humans. His work turned to animal studies rather than humans because of his chance observation of chickens being fed on table scraps in the hospital yard. When polishings were added to the rice leftovers from the table the chickens quickly recovered from the neuritis. Thus the prevention and cure for the dreaded beriberi came about almost by accident. The important fact here is that Eijkman produced a nutritional deficiency experimentally, then overcame it by the use of a supplement. Within two decades after Eijkman's discovery, vitamins were generally recognized to be essential nutrients for humans and animals. The Polish scientist Casimir Funk coined the word *vitamin* in 1911.

The period 1910–1950 was the "Golden Era" of vitamin research. Most of the fat-soluble (A, D, E, K) and water-soluble vitamins (thiamin, riboflavin, niacin, pantothenic acid, and folic acid) were discovered during this period. Another major development was the isolation and chemical identification of vitamins. To Drs. Szent-Györgyi, King, and Waugh belong the honor for isolating and identifying vitamin C as ascorbic acid, the first vitamin to be chemically identified. In more recent times, other vitamins have been isolated and identified. The chemical nature of most of the vitamins has now been established, as has the need for certain fatty acids. Polyunsaturated fatty acids (those with more than one double bond), linoleic, and arachidonic acids have been shown to be required by humans.

Much as also been learned about amino acid requirements during the past 50 years. For example, we now know that there are more than 22 amino acids and that these compounds constitute the main building blocks of proteins. Of these, 9 are essential in human nutrition; that is, they are required for healthy function of the human body and cannot be synthesized in adequate amounts by it. All of the other amino acids can be synthesized in the body from the essential acids and/or from other nitrogen compounds. Still more recent in nutrition history is the recognition of nutritional inhibitors—naturally occurring chemicals in foods that adversely affect the functioning of some nutrients (See Chapter 4).

The end is not yet in sight for completing our scientific knowledge of human nutrition. In fact, we are just beginning to understand the interrelationships among nutrients and the precise role they play in controlling certain physiological disorders of humans (e.g., heart disease, obesity, gastrointestinal disturbances). Then too, we probably have not discovered the last of the vitamins, essential minerals, or inhibitors. It would be fair to say that the simple facts of nutrition have been uncovered, but the more difficult ones are still ahead of us.

Essential Nutrients—Chemistry and Functions

Essential nutrients for humans may be defined as those chemical elements and compounds that are required by them and that must be supplied through the diet or from the environment. These chemicals are needed to satisfy the body's basic physiological needs for growth, maintenance, and repair of tissues and fluids, for manufacture of milk and other fluids, and for reproduction. There are six classes of nutrients: oxygen, water, energy sources, proteins, vitamins, and minerals. Each nutrient group has a specific physiological function, although proteins can serve as sources of both energy and amino acids. Dietary fiber will also be included here, although it is not, strictly speaking, an essential nutrient.

Oxygen

The element oxygen is not commonly regarded as a nutrient, in spite of the fact that without an adequate supply we would be unable to utilize the energy in fats, carbohydrates, and (when necessary) proteins. The need for oxygen is continuous: deprivation for only a few moments brings death. As energy foods are metabolized the oxygen combines with the carbon and hydrogen present to yield water, carbon dioxide and, of course, heat.

Water

Water is of critical importance in nutrition. A person can go without food for a week or more without serious effects, but without water he cannot survive more than a day or two. Water makes up from one-half to three fourths of the weight of the body, and because there are considerable daily losses from the body, water deficiencies can occur rather easily. Even modest water deprivation leads to general body weakness, lassitude, and, of course, thirst; any significant tissue dehydration leads to mental confusion and sullenness. Suffering and death from a lack of water can be quick and horrible, as many reports from the desert and high seas bear witness.

Considering that all of the metabolic reactions of the body are carried out in an aqueous medium, it is no wonder that water is so important. Small changes in the water content of tissues and fluids may profoundly affect the nature and extent of these reactions and thus are the basis for the gross effects of water deprivation just described. The effects are so basic to survival that the regulation of moisture in the body is under very precise physiological control. This control takes place automatically and instantly: when there is a demand for more water by certain tissues, a signal goes out via the nervous system, which triggers the thirst sense and causes a person to consume more water. Urine excretion is the primary means by which water is lost from the body but of course moisture is also expelled during breathing. The amount lost via breathing varies considerably but averages about one third of a quart a day. Water is also lost through the pores in the skin, both as water vapor and as sweat. A limited amount of water is also lost from the body through the saliva and feces.

Energy

The body has a constant demand for energy for numerous physiological processes. With its continuous need for fuel, it can be compared to a machine running 24 hours a day. Just as a car needs gasoline for fuel, so the body requires energy foods—carbohydrate and fat and, if necessary, protein.

Regulation of energy intake, like moisture, is also under precise physiological control. A deficit of energy in the body triggers the sensation of hunger, and this leads to food intake. This reaction is in contrast to that for vitamins and minerals. In the case of these two classes of nutrients even a severe deficiency does not lead to increased appetite for food. In fact, a person can develop a serious deficiency from a lack of minerals or vitamins without feeling hunger!

Energy is needed by the body for carrying out body activities (e.g., exercise, blood circulation, and digestion of food) and for maintaining the body temperature. Energy can be stored in the body for future use, as adipose tissue.

Figure 32 shows percent utilization of food energy by the human. As may be seen, over 50% of the energy is used for basic metabolic functions and 25% for muscular activity. The remainder is used for energy storage, in waste disposal, and for the specific dynamic effect (the increased metabolism associated with ingestion and digestion of food).

Food energy is usually measured in calories. (A calorie is the amount of heat required to raise the temperature of 1000 gm of water 1°C.) The energy contents of nutrient groups as measured in the laboratory using a bomb

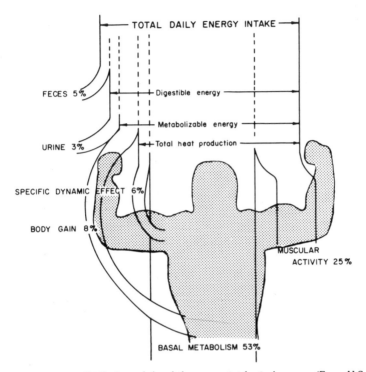

Fig. 32. Average percentage distribution of the daily energy intake in humans. (From U.S. Department of Agriculture, 1959.)

calorimeter* are as follows: carbohydrates, 4.1 calories/gm; fats, 9.5 calories/gm; and protein, 5.7 calories/gm. In the human, digestibility and metabolic utilization effects result in a lowering of the figures for fats and proteins to 9.0 and 4.0 calories, respectively.

Carbohydrates. The most important energy sources in the human diet are the carbohydrates. In the United States carbohydrate foods make up about half of our nutrient intake. In other parts of the world, especially in developing countries, the figure is considerably higher. The nutritionally important carbohydrates are: 1) monosaccharides (e.g., glucose, galactose, and fructose); 2) disaccharides (e.g., sucrose and maltose); and 3) polysaccharides (e.g., starch, dextrin, and glycogen). Ribose, another monosaccharide, is a very

---

*A laboratory instrument that permits measurement of the amount of energy in a food by burning it in oxygen and measuring the amount of heat produced.

important component of many metabolic reactions in the body but is not an important food source of energy. The formulas of some common food carbohydrates are shown here.

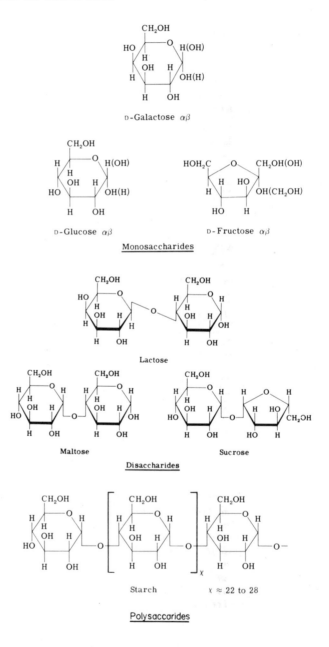

D-Galactose $\alpha\beta$

D-Glucose $\alpha\beta$ D-Fructose $\alpha\beta$

Monosaccharides

Lactose

Maltose   Sucrose

Disaccharides

Starch $\chi \approx 22$ to 28

Polysaccarides

The digestion, absorption, and metabolism of carbohydrate to yield usable energy is in one way simple and another way very complex. Carbohydrates are eventually hydrolyzed to glucose. Glucose is oxidized in the body as follows:

$$C_6H_{12}O_6 + 6\ O_2 \leftrightarrow 6\ CO_2 + 6\ H_2O + \Delta H$$

glucose   oxygen   carbon   water   heat
dioxide

This reaction shows that a molecule of glucose reacts with six of oxygen to yield six molecules each of carbon dioxide and water in addition to heat (4.1 calories/gm of glucose). The exact sequence of the chemical events that take place in glucose metabolism is much more complex. By an elaborate series of enzyme-induced reactions, glucose is broken down to yeild heat or an energy-rich phosphorus compound (adenosine triphosphate, ATP) and carbon dioxide and water (Fig. 33).

Fats. The fats make up a smaller proportion of our energy foods but still are a significant source of calories (about 40% of the total for Americans). Thee most important compounds are triglycerides and phospholipids. The digestion, absorption, and metabolism of fats are not as fully understood as true for carbohydrates; nonetheless, much is known (see Fig. 34). Basically, the

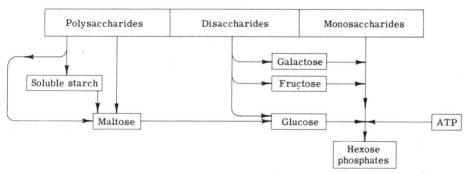

Fig. 33. Pathways of carbohydrate digestion and absorption. The monosaccharides (glucose, galactose, and fructose) are absorbed without breakdown through the intestinal wall into the bloodstream and there are transformed into hexose phosphates. The salivary enzyme amylase converts polysaccharides (e.g., insoluble starch) to soluble starch and also releases some maltose. In the intestine the pancreatic enzyme amylase hydrolyzes the soluble starches, dextrin and glycogen, to maltose. In the case of the disaccharides, the enzyme sucrase splits sucrose to glucose and fructose and the enzyme lactase splits lactose to glucose and galactose. As noted above, these monosaccharides are then transferred through the mucosa into the bloodstream. Here they are transformed into hexose phosphate by interaction with adenosine triphosphate (ATP).

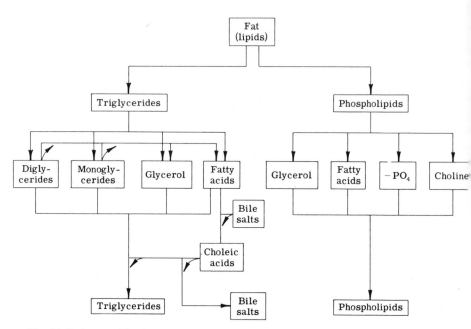

Fig. 34. Pathways of fat digestion and absorption. Lipolytic enzymes, activated by the bile salts in the intestine, attack fats liberating (a) from triglycerides—mono- and diglycerides, glycerol, and fatty acids; and (b) from phospholipids—e.g: for lecithin—glycerol, fatty acids, phosphate, and choline. Triglycerides and phospholipids of different fatty acid compositions are re-formed from these components and absorbed into the bloodstream.

utilization of fat for energy is much like that for carbohydrates: it is oxidized to yield energy, water, and carbon dioxide. For every gram of fat oxidized, 9.0 calories of energy are released for physiological purposes.

Protein (Amino Acids)

Protein—or, to be more exact, certain amino acids contained in it—represents another essential nutrient group. Proteins make up most of the body's structures and almost all of those of the functioning units of the body: hair, skin, flesh, blood, nerves, brain, connective tissues, enzymes, hormones, antibodies, hemoglobin, and myoglobin, among others. The body builds protein from amino acids derived from food as well as from those it synthesizes.

There are more than 20 known amino acids, of which 9 (10 for the very young) must be obtained directly from food. The body can synthesize the remainder of the amino acids needed to manufacture the tissue and fluid proteins of the body.

The structural formulas of nine essential amino acids are shown in Table 18.

All amino acids contain carbon, hydrogen, oxygen, and nitrogen; three also contain sulfur, and two iodine. They all possess an acidic (carboxyl) and an α-amine (or amino) group.

TABLE 18
Structural Formulas for Nine Essential Amino Acids

Essential amino acids	Formula	Molecular weight	Structure	Classification
L-histidine	$C_6H_9N_3O_2$	155.2	$CH_2-CH-COOH$, NH_2 (imidazole ring)	Basic
L-isoleucine	$C_6H_{13}NO_2$	131.2	$C_2H_5-CH-CH-COOH$, CH_3 NH_2	Neutral (aliphatic)
L-leucine	$C_6H_{13}NO_2$	131.2	$(CH_3)_2CH-CH_2-CH-COOH$, NH_2	Neutral (aliphatic)
L-lysine	$C_6H_{14}N_2O_2$	146.2	$H_2N-(CH_2)_4-CH-COOH$, NH_2	Basic
L-methionine	$C_5H_{11}NO_2S$	149.2	$CH_3-S-CH_2-CH-COOH$, NH_2	Neutral (sulfur-containing)
L-phenylalanine	$C_9H_{11}NO_2$	165.2	$CH_2-CH-COOH$, NH_2 (phenyl ring)	Neutral (aromatic)
L-threonine	$C_4H_9NO_3$	119.1	$CH_3-CH-CH-COOH$, OH NH_2	Neutral (aliphatic)
L-tryptophan	$C_{11}H_{12}N_2O_2$	204.2	$CH_2-CH-COOH$, NH_2 (indole ring)	Neutral (aromatic)
L-valine	$C_5H_{11}NO_2$	117.1	$(CH_3)_2CH-CH-COOH$, NH_2	Neutral (aliphatic)

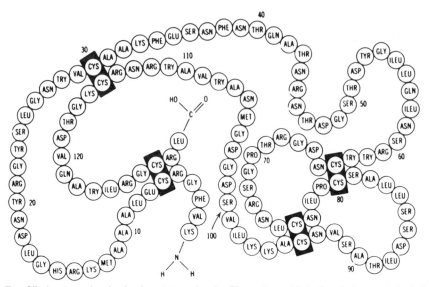

Fig. 35. An example of a food protein molecule. The amino acids in the chain are connected by the peptide bond. In this case the chains are linked with four —S—S— bonds. (From Canfield and Liu, 1965.)

In proteins, the amino acids are bound one to another through the peptide bond. In this bond the α-amino group of one acid is linked to the carboxyl group of the adjacent acid. Proteins are generally long-chain molecules of very high molecular weight (see Fig. 35).

The digestion and absorption steps for protein are shown in Fig. 36. In the body the amino acids undergo a wide variety of reactions and transformations. Of prime importance is the synthesis of tissue and fluid proteins and enzymes, structural elements, and so on. The essential amino acids have special roles because they contain unique chemical structures that the body cannot synthesize. The products of deamination, transamination, decarboxylation, and

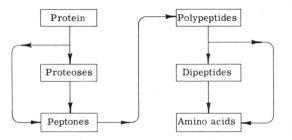

Fig. 36. Pathways of protein digestion and absorption. Protein digestion in the alimentary tract consists of enzymatic cleavage of the protein molecule into smaller and smaller units, down to amino acids. These are absorbed into the bloodstream, where they are transported and synthesized into body proteins.

other metabolic reactions of many of the amino acids are known, as well as the pathways involved. They are extremely complex and varied.

Minerals

The minerals play a very important role in human nutrition. The more common ones, such as calcium and phosphorus, are essential in assuring normal structure and proper functioning of bones and teeth. In fact, over 90% of the calcium and phosphorus in the human body is found in these two structures. The remainder is located in the soft tissue and fluids of the body. Calcium and phosphorus, along with oxygen and hydrogen, are the main elements comprising the complex structure in bone and teeth. These structures are crystalline; their units are organized into a honeycomb-like formation built around a protein matrix. These special compositions and structures provide the rigidity and strength required for the proper functioning of bone and teeth.

Bone building, maintenance, and repair require not only an adequate supply of calcium and phosphorus but also trace minerals and the vitamins A and D. The minerals must be supplied in the diet more or less continuously because only limited amounts can be stored.

The mineral elements know to be required by humans include—in addition to calcium and phosphorus—sodium, potassium, magnesium, copper, iron, manganese, zinc, molybdenum, chromium, nickel, chlorine, iodine, and fluorine. Cobalt is also required but only as part of the vitamin B_{12} molecule.

The specific need for calcium and phosphorus has already been mentioned. Sodium is essential for the functioning of the body fluids and as a small but necessary component of tissues. On the other hand, potassium is mostly concentrated in tissue, with much smaller amounts being found in body fluids. Magnesium is found chiefly in bones, but small amounts are also present in soft tissues and blood.

Sodium and potassium have an important role in controlling water balance among the various tissues and fluids of the body. Evidence is mounting to show that sodium intake, if excessive, is linked to hypertension, a serious disorder in the human. Magnesium serves as a catalyst in a number of metabolic reactions and is an essential part of some of the complex molecules involved in these reactions.

Iodine is essential for the synthesis of thyroxin, the hormone that controls the rate of many physiological functions. Lack of iodine results in goiter (Fig. 37). Iron and copper are required for the formation and functioning of hemoglobin and myoglobin, the oxygen-transporting and storage compounds of blood and muscle. These elements are also essential parts of certain enzyme systems. Fluorine functions in the body in protecting teeth against caries (i.e., tooth decay).

Fig. 37. Goiter from iodine deficiency. (Courtesy Dr. George Briggs, University of California at Berkeley.)

Vitamins

The chemicals known as vitamins make up a sizable group of nutrients that have widely varying chemistries and perform some very important functions in the metabolism and utilization of carbohydrates, fatty acids, and amino acids in the diet (see Table 19). In this section we will examine their chemical nature and functions in human nutrition. (In passing, it is interesting and important to note that not all of the higher animals require the same vitamins. For example, the chicken does not need vitamin C in its diet; the bird is capable of synthesizing it in the body.)

Vitamins are classified as fat-soluble and water-soluble.

Fat-Soluble Vitamins (Fig. 38). *Vitamin A* (retinol) is a complex unsaturated cyclic alcohol, that exists as several isomers. These have differing biological activities, with the all-trans forms possessing the highest activities.

Two derivatives of vitamin A, the acetic and palmitic acid esters, are synthesized and sold commercially for the nutrient enrichment of certain foods. (See the last section of this chapter, Nutritional Enrichment of Food)

Vitamin A is not found in plants as such; however, it can be readily synthesized in the body from certain carotenoids (the yellow-orange-red

TABLE 19
The Vitamins—Physiological Functions and Deficiency Symptoms[a]

Vitamin	Physiological functions	Deficiency symptoms
Fat-soluble		
Vitamin A (retinol)	Rod vision, growth factor, maintenance of epithelial integrity	Xerophthalmia, keratomalacia, blindness
Vitamin D (calciferol)	Calcium and phosphate metabolism	Rickets, osteomalacia
Vitamin E (tocols and trienols)	Biological antioxidant, intracellular respiration, cellular and vascular integrity, central nervous system and muscle integrity	Microcytic anemia and edema; creatinuria, red cell hemolysis, ceroid pigment
Vitamin K (phylloquinones)	Formation of certain blood-clotting factors	Hemorrhage, decreased clotting
Water-soluble		
Thiamin (B_1)	Aldehyde transfer, metabolism of carbohydrates	Beriberi
Riboflavin (B_2)	Hydrogen and electron transfer, role in metabolism of macronutrients	Glossitis, dermatitis, cheilosis, ocular symptoms
Niacin (nicotinic acid)	Hydrogen transfer, degradation and synthesis of fatty acids, carbohydrates, and amino acids	Pellagra
Folacin (folic acid)	Formyl transfer, C_1 metabolism	Macrocytic anemia
Biotin	CO_2 transfer, carboxyl group transfer, fatty acid biosynthesis	Seborrheic dermatitis
Pantothenic acid	Acyl transfer, role in macronutrient metabolism	Gastrointestinal and nervous disorders
Pyridoxine (B_6)	NH_2 transfer, other functions in amino acid metabolism	Convulsions, dermatitis
Cobalamine (B_{12}, cyanocobalamine)	Dehydrogenation, methylation	Pernicious anemia
Vitamin C (ascorbic acid)	Integrity of intercellular substances, oxidation—reduction systems	Scurvy
Choline	Fat metabolism, component of phospholipids, methyl donor for transmethylation	Fatty liver

[a]Adapted from Peterson and Johnson (1978).

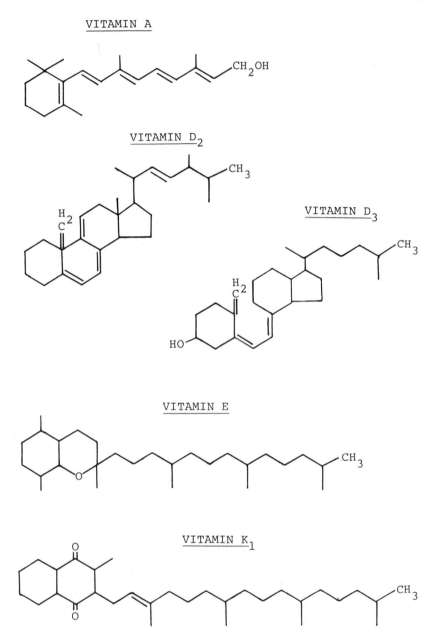

Fig. 38. Chemical structures of the fat-soluble vitamins.

pigments) present in many plant food raw materials. Vitamin A is found in many animal food raw materials, such as milk fat, egg yolk, and fish oils. This essential nutrient plays a very important and specific role in vision (see Table 19 and Fig. 39), in growth, in reproduction, in the normal functioning of certain mucus-secreting cells of the body, and in other as yet unspecified ways. Excessive dietary levels of vitamin A are toxic to humans. At very high levels (10 times the recommended daily allowance), it produces severe joint pains and even loss of hair.

Vitamin D (also termed calciferol) is another complex sterol that occurs in two different molecular forms, both of which are biologically active in humans. The D_3 form is produced by the action of ultraviolet (UV) radiation (from sunlight or a UV lamp) on 7-dehydrocholesterol, normally found in human skin. The D_2 form is produced *in vitro* by the action of UV radiation on ergosterol, a plant sterol; D_2 is synthesized commercially and is in common use to fortify beverage milk. There may be other forms of vitamin D, according to some very recent reports.

This vitamin plays a key role in the mineralization (incorporation of calcium and phosphorus) of bones and teeth, thus providing rigidity and strength to these structures. Vitamin D is toxic to humans when consumed at high levels. Its toxicity is expressed in excessive calcification and hardening of certain soft tissues of the body, especially in the kidney, heart, arteries, and stomach. Infants are especially vulnerable to this disorder.

Vitamin E is one of a group of organic compounds called tocols and trienols, of which α-tocopherol, has the highest biological activity. It is prepared commercially from plant sources and is used to some extent in the nutrient en-

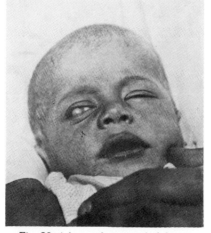

Fig. 39. Advanced vitamin A deficiency.

ASCORBIC ACID

CYANOCOBALAMIN

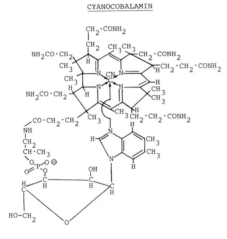

BIOTIN

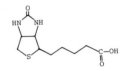

CHOLINE

$(CH_3)_3NCH_2CH_2OH$
              |
              OH

FOLACIN

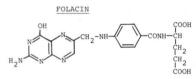

NIACIN

NIACINAMIDE

PANTOTHENIC ACID

PYRIDOXINE

RIBOFLAVIN

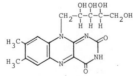

THIAMINE HYDROCHLORIDE

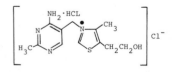

Fig. 40. Chemical structures of the water-soluble vitamins.

richment of food. Also, interestingly enough, it finds use as an antioxidant in food preservation (see Chapter 8).

The precise roles of vitamin E in human nutrition remain unclear, despite considerable research effort. It is agreed, however, that this nutrient serves as an antioxidant in body tissues and fluids, thus helping to prevent the formation of toxic oxidation products during the metabolism and utilization of other nutrients. In addition, it may play a role in certain important enzyme reactions. Much more research is needed on vitamin E and fortunately studies are currently under way around the world. Vitamin E does not appear to be toxic at intake levels associated with food.

Vitamin K represents a group of complex quinone compounds, all of which possess some biological activity. These compounds are widely distributed in nature as components of microorganisms and of plant and animal tissues. Vitamin K is essential in the maintenance of an effective blood-clotting system in the body. It is specifically involved in the synthesis of four blood-clotting proteins. This nutrient is nontoxic at intake levels associated with food.

Water-Soluble Vitamins (Fig. 40). *Thiamin* (B_1) is a complex molecule consisting of a pyrimidine and a unique thiazole nucleus joined together by a methylene bridge. It is chemically synthesized commercially (as the hydrochloride and the mononitrate) and is used for the nutrient enrichment of food. Thiamin is one of the B vitamins, which are essential in the metabolism and utilization of carbohydrates in the body, eventually leading to the production of energy for the maintenance of body temperature, for muscular activity, and for storage as fat (Fig. 41). Thiamin is nontoxic at intake levels associated with food.

Riboflavin (also called B_2) is a complex aromatic compound, with an isoalloxazine nucleus ring with a ribose side chain. The phosphate salt of riboflavin (which is water soluble), as well as the natural chemical form, are synthesized commercially and used for the nutrient enrichment of food. Riboflavin, too, plays a key role in energy metabolism in the body, involving the utilization of carbohydrates, fatty acids, and amino acids. This vitamin is nontoxic at intake levels associated with food.

Niacin (nicotinic acid) is a relatively simple aromatic molecule, made up of a pyridine ring with a carboxyl side chain. The vitamin is available commercially for food fortification. Niacin is another of the B vitamins involved in the metabolism and utilization of carbohydrates, fatty acids, and amino acids in the body. It is interesting to note that the amino acid tryptophan is a biological precursor of niacin. This vitamin is nontoxic when consumed at levels normally found in food.

Vitamin B_6 includes Three compounds (pyridoxine, pyridoxal, and pyridoxamine), all of which are biologically active. These are fairly simple organic

Fig. 41. Beriberi. (From Vedder, 1913).

compounds made up of a hydroxypyridine ring with an alcohol, aldehyde, or amine side chain. The form with the alcohol side chain is synthesized chemically and is sold commercially for food fortification. More than 60 enzymes, many of which are involved in the metabolism and utilization of carbohydrates and essential fatty acids in the body, depend on derivatives of B_6 for their activities. Vitamin B_6 has other functions as well. It is nontoxic when consumed at levels normally found in food.

Pantothenic acid is a relatively simple aliphatic acid, only one isomer (dextro-) of which is biologically active. The calcium salt is available commercially for the enrichment of food. This vitamin plays a key role in the metabolism of carbohydrates, fatty acids, and some of the amino acids in the body. It is of importance in the synthesis of certain important compounds, including steroids (which include certain hormones) and acetylcholine, the vital chemical transmitter of impulses originating at nerve endings. It also has other functions.

Pantothenic acid is nontoxic when consumed in amounts normally found in foods.

Biotin is a somewhat complex compound made up of a monocarboxylic acid connected to a cyclic urea ring containing sulfur. There are eight isomers of this molecule, only one (dextro-) of which is found in nature. It is biologically active. Biotin functions in intermediary metabolism and is associated with certain carboxylation reactions, including the synthesis of long-chain fatty acids in the body. Biotin is nontoxic at intake levels associated with food.

Folic acid (also called folacin) is a complex organic molecule that contains the unique pteridine ring linked to a glutamic acid group. In nature the molecule may contain one glutamic acid group, but more commonly the ring is conjugated with two or more glutamic acid groups. The molecular form containing one glutamic acid group is available commercially and is used in food fortification. This vitamin is nontoxic when consumed in amounts usually found in food.

Vitamin B_{12} (also called cyanocobalamine or cobalamine) is the most chemically complex of all the vitamins. The organic moieties of the molecule surround a single centrally located cobalt atom. Its chemical synthesis was not accomplished until 1974. In nature, B_{12} is synthesized mostly by microorganisms and becomes part of food raw materials only indirectly, as a microbial contaminant in the case of plants and as a by-product of microbial activity in the gut in the case of other animals. Vitamin B_{12} is essential in intermediary metabolism and is involved in the reduction of certain organic compounds that, with folic acid, are involved in the synthesis of labile methyl groups. The latter are essential for the formation of the important purine and pyrimidine bases. This vitamin also plays other roles in the metabolism and utilization of carbohydrates and amino acids. Vitamin B_{12} is largely associated with foods of animal origin. This fact presents a special problem for strict vegetarians, but ovo-lacto vegetarians naturally do not have such a problem. Vitamin B_{12} is nontoxic when consumed in amounts normally found in food.

Vitamin C (ascorbic acid) is a simple aliphatic compound: a hexuronic acid. It exists in nature in both the reduced and the oxidized forms (the latter as dehydro-ascorbic acid), both of which are biologically active. The vitamin is chemically synthesized and is sold commercially for food fortification. Interestingly enough, it also is used as an antioxidant in food preservation (see Chapters 7 and 8). Ascorbic acid is essential for the normal functioning of both cellular and subcellular units in all plant species and in higher animals. It functions specifically in the oxidation of certain amino acids and in the hydroxylation of one specific amino acid, proline. It also has other functions in the body. Vitamin C is nontoxic at intake levels normally associated with food.

Choline is a rather simple aliphatic compound (β-hydroxyethyl) trimethyl ammonium hydroxide. It is a basic constituent of the common phospholipid, lecithin. It is essential in the synthesis of phosphatidylcholine found in cell membranes and in lipoproteins, which are involved in the important role of transporting fat-soluble compounds in the body. It is nontoxic at intake levels associated with food.

Essential Fatty Acids

There are three polyunsaturated fatty acids of interest in human nutrition that are considered to be essential: linoleic, linolenic, and arachidonic. However, because linolenic can be synthesized by the body, linoleic and arachidonic acids are the true essential fatty acids. These compounds are rather simple aliphatic polyunsaturated acids containing the methylene-interrupted double bond (—C=C—C=C—). Only the cis—cis isomers are biologically active. These fatty acids are involved in the maintenance of membrane structure and function in the body. In addition, they serve as precursor compounds in the synthesis of so-called internal hormones, which have diverse functions in metabolism, in nerve function, and in certain secretions of the body. There is some controversy concerning the toxicity of high-level intakes of polyunsaturated fats, but some evidence is accumulating to indicate that there may be cause for concern. More research is needed and much work on the problem is in progress around the world.

Dietary Fiber

In recent years we have become aware of the importance of fiber in the diet. Dietary fiber has been defined as those remnants of food plant cells that resist digestion and absorption in the gastrointestinal tract. These remnants may consist of one or more of the following compounds: cellulose, hemicellulose, pectin, lignin, and certain undigested proteins. These substances bind considerable water and thus increase fecal bulk, which in turn reduces the transit time of the bowel contents through the gut. The positive effect is to reduce constipation and any tendency toward diverticulosis (a serious structural disorder of the large intestine).

Human Nutritional Requirements

A great deal of time and energy has been spent by highly qualified nutritionists in establishing the quantitative requirements of essential nutrients for human beings. In attempting to obtain reasonably valid estimates they have examined published data based on limited human studies, and also data from epidemiological surveys relating nutrient intake to the incidence of nutritional deficiencies diseases in a number of countries.* In addition, these experts have examined the voluminous data gathered from experimental animal studies and controlled human studies.

Progress in obtaining sufficient, reliable, and relevant data has been slow for several reasons. Comparatively few human studies have thus far been conducted, and these have involved only a limited number of nutrients. The epidemiological studies, though quite numerous and often carefully done, provide only indirect evidence of requirements. These surveys permit the establishment of a correlation between nutrient intakes and deficiencies, not a cause-and-effect relationship. Still they do yield valuable clues about requirements. Experimental animal studies, especially those using species that require the same essential nutrients as humans and that also metabolize and utilize them in similar ways, also provide valuable data.

As a result nutritionists in the United States and around the world have developed what are generally regarded as fairly reliable estimates of the quantitative daily human requirements for the essential nutrients. The current status of some of these efforts will be discussed next.

FAO/WHO Standards

The Food and Agriculture Organization (FAO) and the World Health Organization (WHO), both agencies of the United Nations, have a joint working group that sets standards for human nutritional requirements. These are provided by FAO and WHO especially for use in a variety of programs sponsored by or carried out by them in countries around the world that have nutritional problems (see Chapter 2). FAO/WHO recommendations (Beaton and Patwardham, 1976) provide daily intake levels for 11 essential nutrients plus energy for 12 age/sex categories as well as for pregnancy and lactation. Requirements for the remaining essential nutrients are expected to be found in sufficient amounts in the traditional foods consumed by people in the various parts of the world. The publication containing these standards also briefly discusses the need for iodine and fluorine because of their known importance and their general unavailability in certain areas of the world. In addition, there is a short discussion of the need for such trace metals as zinc, magnesium, copper, chromium, and molybdenum for similar reasons.

NRC/NAS Recommended Daily Allowances (RDAs)

Similar recommendations are published periodically by the U.S. National Research Council/National Academy of Sciences (NRC/NAS) as an official

---

*Epidemiological studies are made on selected populations and correlate relationships between, in this case, incidence of nutritional deficiencies and food habits.

document prepared by the NRC Food and Nutrition Board. In the published document recommendations are made for 17 essential nutrients plus energy for 15 age/sex categories as well as for pregnancy and lactation (see Table 20 for examples). Short discussions relevant to meeting these nutritional requirements are also included.

The recommendations of FAO/WHO and NRC/NAS differ somewhat, as might be expected, considering the incomplete state of our knowledge of the subject and the fact that different nutritionists are involved in developing the recommendations. Also, the FAO/WHO figures are developed for worldwide application, whereas those of the NRC/NAS are for application in the United States.

The RDAs provide a margin of safety above minimum requirements for protein, minerals, vitamins, and energy. This margin varies for the different nutrients, depending on the ability of the body to store the nutrient, natural variation among individuals, and possible toxicity effects, among other factors. These allowances are based on the amount of the nutrients found in food as consumed and not on the levels found in food before processing, storage, distribution, preparation, and service to consumers. In other words, all of the losses that occur before consumption must be taken into account in figuring diets that meet these allowances. On the other hand, these allowances do recognize the lack of physiological availability or incomplete absorption from the intestinal tract of the nutrients in certain foods. Older people are generally very much less active than young persons and therefore need fewer calories. Calorie allowances, therefore, are adjusted for body weight, age, and physical activity. Then too, there are persons whose requirements are larger or smaller than average; therefore, adjustments are made in calculating one's energy requirements.

Diets that supply less than the recommended quantitative allowances may not necessarily lead to deficiency symptoms. Individuals differ greatly in their specific requirements as well as in their ability to adjust to diets that supply less than what might appear to be adequate nutrients. The official dietary allowances have been set high enough to care for those whose requirements are higher than average. Accordingly there is a considerable margin of safety in these allowances for a majority of people.

There are a number of nutrients that are recognized as being essential for humans but for which official allowances have not been specified by the Food and Nutrition Board. These include specific amounts of carbohydrate and fat, water, several minerals (sodium, potassium, copper, cobalt, zinc, manganese, and molybdenum), and a number of the vitamins (e.g., biotin and vitamin K). The board has not set any allowances for these nutrients, in part because requirements for some of them are not known and in part because deficiencies are not likely to occur in diets commonly in use in the United States. In other words, a diet that supplies the recommended allowances will probably also

TABLE 20
Estimated Safe and Adequate Daily Dietary Intake Levels of Additional Selected Vitamins and Minerals[a,b]

	Age (years)	Vitamins			Trace elements[c]						Electrolytes		
		Vitamin K (μg)	Biotin (μg)	Pantothenic acid (mg)	Copper (mg)	Manganese (mg)	Fluoride (mg)	Chromium (mg)	Selenium (mg)	Molybdenum (mg)	Sodium (mg)	Potassium (mg)	Chloride (mg)
Infants	0 – 0.5	12	35	2	0.5–0.7	0.5–0.7	0.1–0.5	0.01–0.04	0.01–0.04	0.03–0.06	115–350	350–925	275–700
	0.5– 1	10–20	50	3	0.7–1.0	0.7–1.0	0.2–1.0	0.02–0.06	0.02–0.06	0.04–0.08	250–750	425–1275	400–1200
Children	1 – 3	15–30	65	3	1.0–1.5	1.0–1.5	0.5–1.5	0.02–0.08	0.02–0.08	0.05–0.1	325–975	550–1650	500–1500
and	4 – 6	20–40	85	3–4	1.5–2.0	1.5–2.0	1.0–2.5	0.03–0.12	0.03–0.12	0.06–0.15	450–1350	775–2325	700–2100
adolescents	7 –10	30–60	120	4–5	2.0–2.5	2.0–3.0	1.5–2.5	0.05–0.2	0.05–0.2	0.1 –0.3	600–1800	1000–3000	925–2775
	11 +	50–100	100–200	4–7	2.0–3.0	2.5–5.0	1.5–2.5	0.05–0.2	0.05–0.2	0.15–0.5	900–2700	1525–4575	1400–4200
Adults		70–140	100–200	4–7	2.0–3.0	2.5–5.0	1.5–4.0	0.05–0.2	0.05–0.2	0.15–0.5	1100–3300	1875–5625	1700–5100

[a]Reproduced from Food and Nutrition Board (1980).
[b]Because of inadequate information on which to base allowances these figures are provided in the form of ranges of recommended intakes.
[c]Because of toxic levels for many trace elements may be only several times usual intakes, the upper levels for the trace elements given in this table should not be habitually exceeded.

supply enough of the other nutrients to meet requirements. It is well recognized that foods contain many essential nutrients in addition to the ones for which there are recommended allowances.

Nutrient Composition of Food

The nutrient composition of common foods in the American diet has been studied extensively. However, much of the information is based on inadequate methodology or on too few samples; and in some cases the published figures are questionable (Stewart, 1979).

Differences in the kind and amount of chief nutrients among the various foods are readily apparent from these studies. Milk, for example, is noted for its calcium content but contains little iron. Meat provides considerable protein but only a negligible amount of calcium. Oranges are a superior source of ascorbic acid but contain almost no protein. Similarities in nutrient content exist among many of our foods. Meat, fish, and poultry, for example, are excellent sources of protein, as are eggs, milk, dry beans and peas, and nuts. Similarly, green and yellow vegetables are good sources of provitamin A (carotene), and citrus, tomatoes, strawberries, and cabbage are valuable for their ascorbic acid (vitamin C) content. Whole-grain cereals and cereals enriched with vitamins and minerals provide, in addition to calories, substantial quantities of iron, thiamin, riboflavin, and niacin.

Food Groups

Following are a few examples of the natural groups of foods on the basis of similarity of nutrient composition. These lend themselves to a variety of uses, such as in the appraisal of individual diets. It is possible to set up 11 such groups.

1. Milk and milk products. This group includes milk (whole, skim, and buttermilk), yogurt, cheese, cream, and ice cream. The amounts suggested for diets and menus are generally given in terms of quarts of fluid milk per day. For example, when using cheese or ice cream its milk equivalent is calculated as follows: 1 pound of cheddar-type cheese is equivalent to 3 quarts of milk; a 4-ounce package of cream cheese to one-fourth cup of milk; a 12-ounce container of cottage cheese to about 1 cup of milk; and 1 quart of ice cream to 1 pint of milk.

2. Meat, poultry, and fish. This group includes beef, veal, lamb, pork; poultry, fish, and shellfish. In suggesting the amounts of meat for diets and menus it is assumed that no more than one-third pound of bacon and salt pork is being used for each 5 pounds of other meat, poultry, and fish. Other-

wise the high fat content of these cuts results in a decrease in protein intake and an increase in energy.

3. Eggs. Fried, boiled, or poached eggs, or eggs used in cooking are included in this group.

4. Dry beans, peas, and nuts. This group comprises dry beans of all kinds, dry peas, lentils, other beans, and oil seeds such as soybeans, peanuts, and tree nuts and their products.

5. Flour, cereals, and baked goods. This group includes wheat flour and cornmeal, cereals (including ready-to-eat products such as rice, hominy, pastas, noodles, macaroni, and spaghetti), tacos, grits, bread, cake, and other baked goods. The amounts suggested for diets and menus are in terms of pounds of flour or cereal. Bread and other baked goods average two-thirds flour by weight. Therefore, 1 pound of bread and baked goods is counted as two-thirds of a pound of flour.

6. Citrus and tomatoes. This group comprises grapefruit, lemons, limes, oranges, tangerines, and tomatoes.

7. Dark green and deep yellow vegetables. This group includes spinach, carrots, pumpkin, and yellow winter squash.

8. Potatoes. This group comprises white potatoes, yams, and sweet potatoes.

9. Other vegetables and fruits. This group comprises asparagus, beets, brussels sprouts, cabbage, cauliflower, celery, corn, cucumbers, green lima beans, snap beans, lettuce, okra, onions, peas, rutabagas, sauerkraut, summer squash, turnips; apples, bananas, berries, dates, figs, grapes, peaches, pears, plums, prunes, raisins, rhubarb; and all other vegetables and fruits not included in other groups.

10. Fats and oils. In this group we find butter, margarine, mayonnaise, salad dressing, salad and cooking oils, shortening, and lard.

11. Sugar, syrup, and preserves. This group embraces sugar, maple syrup, molasses, corn syrup, honey, jam, jelly, and preserves.

Variability in Nutritive Values

Variations in the nutrient content of each type of food are common. For example, the protein content of one lot of wheat can be almost double that of another; the fat content of milk depends on breed of cow, stage of lactation, and other factors; and the provitamin A (carotene) content of carrots varies with variety, cultural practices, and other factors. Careful account must be taken of the characteristics of each food in providing average or representative nutritive values. Oranges and potatoes are good examples of the complexity of obtaining meaningful values for individual foods. Oranges are especially high in ascorbic acid (vitamin C), yet orange-to-orange variation is very large.

It has been found that 100 gm of orange juice may contain from less than 20 to more than 80 mg of ascorbic acid. Differences in variety account for much of this difference. For example, the California Navel orange contains about 60 mg/100 gm, whereas juice from California Valencias and Florida-grown early and midseason varieties averages about 51 mg and Florida late-season Valencias may have as little as 37 mg. Vitamin C content of orange juice also varies with the maturity of the fruit, being higher in the early part of the harvest season. Measurements have been taken at different times during the harvest season and for different varieties in calculating an average value. The year-round, countrywide average for fresh orange juice is about 50 mg of ascorbic acid per 100 gm. Commercial frozen orange juice is prepared mostly from Florida fruit and the ascorbic acid content averages about 45 mg per 100 gm. The loss of ascorbic acid during the processing and storage of frozen citrus juices is negligible.

Grapefruit, in contrast to oranges, shows few varietal differences in ascorbic acid content. However, it does show a downward trend seasonally. Juice from fruit harvested in September and October averages about 42–47 mg/100 gm; the lowest values average 33–35 mg and are for spring-harvested fruit.

Potatoes present a different picture. They are moderately high in ascorbic acid but values vary widely. One potato may have more than 50 mg/100 gm, whereas others contain less than 5 mg. The variety, harvest maturity, and storage conditions all influence the ascorbic acid content. Values are highest for immature potatoes (sometimes called "new" potatoes), averaging about 35 mg/100 gm. Few of these potatoes get into commercial channels. Freshly harvested mature potatoes (i.e., harvested potatoes that have not been stored) average about 26 mg of ascorbic acid per 100 gm. Most of the potatoes sold during the winter and spring months have been cold-stored for some time. The ascorbic acid content of such potatoes drops progressively with time: about one fourth of the vitamin C content is lost during the first month; one half after 3 months; and two thirds after 6 months. The year-round value for commercial potatoes, taking into account variety and storage, is 20 mg/100 gm. Ascorbic acid content for the different varieties of potatoes of major commercial importance ranges from 19 to 33 mg/100 gm.

Milk and Milk Products. The nutrient values of milk and dairy products may be summarized as follows: the average composition of milk is about 87% water, 3.5% protein, 4% fat, and 5% carbohydrate. In addition, milk is an excellent source of calcium and a good source of vitamin A, thiamin, riboflavin, and vitamin B_{12}. Skim (nonfat) milk, both fluid and powdered, has become an increasingly common article of commerce. Nearly all the fat has been removed from this product; accordingly, its energy value is reduced greatly from that of whole milk. In fact, a glass of skim milk has only half the calorie content of a glass of whole milk. Removing the fat also removes many of the other nutrients, especially vitamins A, D, E, and K. A pint of whole milk provides nearly a sixth of the entire daily requirement of vitamin A for an adult, whereas a pint of skim milk has only a negligible amount. It is, however, becoming a common practice to add back vitamins A and D to skim milk to make up for the losses. All of the water-soluble nutrients—the minerals, the B vitamins, and the protein—remain in the skim milk.

Milk is the raw material for many manufactured products, especially butter, cheese, and ice cream. The nutrient content of such products depends on whether the raw material used is whole milk or one of the fractions (cream, skim milk, casein, whey, etc.). Either a decrease or an increase in nutrients may result from the addition of the other ingredients in the manufactured product. Cheeses are made by using an enzyme (rennin) and a microbial culture and following a number of processing steps. The first step is to form a curd, most of which is the protein casein. The aqueous portion remaining is whey. After the curd is formed, it is processed in various ways and into different forms. The finished cheese may be aged to produce such types as cheddar, Roquefort, or Swiss. Thus composition and the nutritive value of cheese depend on the raw material used (type of milk) and on the manufacturing steps used. Cheese curd carries with it most of the protein, much of the calcium, and some riboflavin. The nutrients carried off in the whey are made up mostly of lactose, some protein, riboflavin, other B vitamins, and minerals.

Cottage cheese is a soft uncured product made by the action of rennin and a bacterial culture on skim milk, followed by cutting, whey removal, and other processing steps. Cream may be added to the curd to produce a creamed cottage cheese. Uncreamed cottage cheese has about 0.5% fat; creamed cottage cheese possesses a smaller proportion of calcium compared to other types of cheese. The hard whole-milk cheeses (e.g., cheddar, Swiss, Parmesan) have a much higher content of fat and vitamin A than those made from skim milk. Processed cheeses differ from regular cheeses in that they contain additional whey, skim milk, milk, or cream, as well as certain additives to control texture and improve shelf life.

Eggs. The protein content of eggs is relatively constant, but the diet of the hens has a significant influence on the vitamin content of the egg. Nutrients are unevenly distributed among the egg components. The yolk, representing a little more than a third of the contents, contains all the fat, vitamin A, and thiamin, most of the calcium, phosphorus, and iron, and a substantial portion of the protein and riboflavin.

Meat, Poultry, and Fish. Meat, poultry, and fish are valuable sources of nutrients and contain large quantities of high-quality protein, iron, and several of the B vitamins. These nutrients are found mainly in the lean portion, the fat being primarily a good source of energy. Increasing evidence reveals that much of the iron in these products is found bound to heme, a chemical moiety in hemoglobin and myoglobin, the pigments of blood and muscle, respectively. The bound form is more readily absorbed and utilized by humans than other forms of iron.

Beef round steak is one of the leanest cuts, with the lean of the lower grades ranging down to 2% fat. This value increases to 4–5% in the choice ground round. Rib and loin cuts usually have more fat. The lean of these cuts has a high fat content because of fat marbling (fat layered between muscle fibers). Cuts from the loin (e.g., T-bone and club steaks) average 8–10% fat in the loin portion of "Good" grade beef, compared to 12–15% from "Prime" grade. (Good, Choice, and Prime are official terms used to denote the different grades of beef found in the retail market, in increasing order of quality.) The water, protein, and mineral content of beef are not subject to much variation (other than that due to variations in fat content); water makes up about 77%, protein 22%, and minerals 1%. The ratio of water : protein : ash in the lean remains practically the same.

Pork has a total fat content of 4–10% in the lean portion, such as ham, and 6–18% in a cut such as loin. It is important to remember that pork has a much higher content of thiamin than other red meats. Lamb and veal are somewhat different from beef or pork in composition. Veal has more lean and less fat than any other red meat. Organ meats, such as liver and heart, are distinctly different in nutrient composition from muscle. Liver, for example, is an excellent source of vitamins A, C, and B_{12}, and of iron.

As served, red meats consist of the lean and a certain amount of visible fat, plus bone. Many people trim off the fat and eat only the lean; others enjoy the fat or at least a part of it. Accordingly, red meat as eaten varies considerably in composition. The fatty tissue contains a small amount of protein but of course is mainly fat (60–80%). Cooked meat has less water than uncooked and of course well done is less moist than rare or medium done. Accordingly, nutrients become more concentrated during cooking. Lean roast beef, for example, cooked rare contains approximately 3% fat, 21% protein, and 75% water; when cooked to medium doneness it contains about 5% fat, 29% protein, and 65% water.

Broiler-fryers comprise the largest proportion of the commercially available chicken. They have less fat and therefore are lower in caloric value than most other meats, although recently the fat content of commercial broiler-fryers has increased. Most of this fat occurs in deposits in the abdominal cavity and under the skin. The white meat of chicken has a lower fat content than the dark and is also lower in iron, thiamin, and riboflavin. However, it is higher in niacin than dark meat. Turkeys are generally more mature than chickens when marketed and accordingly have a higher fat content. Young turkeys (fryer-roasters), which have become increasingly popular, are lower in fat. The mineral and vitamin content of turkey meat is similar to that of chicken.

The fat content of fish and shellfish varies widely depending on the species and season. Compared with red meat, fish and shellfish are low in fat; even the fatty types rarely run more than 10% and many contain less than 1% fat. The skin of fish is higher in fat than is the flesh. The common method of cooking that employs fat or oil, and the canning of fish in oil, naturally increases fat content. The vitamin content of fish is similar to that of beef, except that some of the fatty fish contain large amounts of vitamin A.

Fruits and Vegetables. Fresh fruits and vegetables are relatively high in moisture, and several of the more succulent ones such as tomatoes, celery, and lettuce contain even more water than orange juice or milk. On the other hand, certain vegetables such as sweet potatoes and green lima beans have a lower moisture content, but even these are made up of about two-thirds water. Peas and potatoes contain about 75% water. The water, mineral, and vitamin content of fruits and vegetables generally changes as maturity approaches. Peas, as an example, decrease in moisture content from 81% at the early stages of maturity to 76% at harvest time. They may even drop to as low as 65%, but at this moisture level they are well below prime quality from the standpoint of flavor and texture. Generally speaking, fruits and vegetables contain very little protein and only traces of fat. However, green peas and lima beans are exceptions to this rule because they may contain as much as 6–8% protein. The carbohydrate content of fruit and vegetables ranges from less than 3% in lettuce to about 23% for bananas. The carbohydrate consists mainly of sugars and starches, but fiber and related carbohydrates are also present, and these constitute most of the dietary fiber mentioned earlier.

The dark green and deep yellow vegetables are rich sources of vitamins A and C; the depth of color is a fairly good index of carotene (provitamin A) content. Accordingly, spinach and broccoli have much more vitamin A value than lettuce, cabbage, or snap beans, and leaf lettuce has much more carotene than the head type. Carrots are also a very rich source of carotene, the amount increasing with maturity. Large carrots usually have a much deeper color and consequently a much higher vitamin A value than small ones. In addition to high levels of vitamin C being associated with citrus fruits and tomatoes, there are substantial amounts of this vitamin in many other fresh fruits and vegetables, for example, cauliflower, cantaloupe, strawberries, cabbage, brussels sprouts, and sweet potatoes. Apples also are a good source, depending, however, on variety and season of maturity, and whether they have been stored or peeled. For example, a large summer apple furnishes about 22 mg of vitamin C per 100 gm if eaten whole but only 14 mg if eaten peeled. Vitamin C content of apples may drop as much as 50% after long storage (by late spring). Certain vegetables are fairly good sources of thiamin. For example, lima beans, peas, and corn contain significant quantities of this vitamin.

Dark green leafy vegetables are also good sources of iron and calcium; however, these minerals are not uniformly distributed in the plants. For example, the calcium content of the outside leaves of head lettuce is three to five times as high as in the pale inner ones. Some greens, such as spinach, contain appreciable amounts of calcium.

Dry Beans, Peas, and Nuts. Mature dry seeds used as food in the United States include navy, pinto, and lima beans; dry peas; and peanuts. These, along with the tree nuts (walnuts, almonds, pecans, etc.), have a low moisture content and are high in fat. Thus they are quite concentrated, nutritionally speaking. They are the richest sources of protein of our plant foods and are also good sources of thiamin. Peanuts are an extraordinarily good source of niacin; the thin reddish-brown skin is especially rich in thiamin. The fat content of nuts averages about 50% or more. These fats contain large amounts of the essential fatty acid linoleic acid.

Flour, Cereals, and Baked Goods. Wheat, corn, rice, and oats and their products are common in American diets. The grains have many nutritional characteristics in common: they are concentrated foods with moisture contents ranging from 15% down to 5%. The cereals vary somewhat in protein content; for example, wheat ranges from 9 to 14%; oatmeal contains about 14%; and rice, about 7%. All cereals are high in carbohydrate, and the endosperm that makes up the major part of the kernel is almost entirely starch. The outside layers of grains contain a considerable amount of dietary fiber. Minerals and vitamins are present in grains, especially in the germ and outer layers of the kernel. The small amount of fat in cereals is concentrated in the germ.

Pure whole-wheat flour contains the germ as well as the outer layers of the kernel and thus possesses all of the nutrients in the original grain. White flour is produced by a milling process that removes the germ and most of the outer layers of the kernel. White flour has a higher proportion of starch and a higher caloric value than whole-wheat flour, but much of the vitamin and mineral content of the kernel is lost in milling white flours. (See Cereal Products, in the next section.)

Prepared cereal products are commonly eaten for breakfast in the United States. Some are made from a single grain, but frequently they are composed of mixtures of refined and processed products made from two or more cereals. These items may contain added nutrients and are then usually called "enriched" or "restored." (See the section on Nutritional Enrichment of Food in this chapter.)

Fats and Oils. Fats and oils include fats derived from milk and meat, as well as oils extracted from vegetable products, fruits, nuts, and grain. Butter, margarine, and salad dressings are also included in this group. Fats and oils are frequently described as being of vegetable or animal origin. Those that are liquid at room temperature are called oils. Oils may be hardened commercially by hydrogenation, that is, by adding hydrogen to the unsaturated bonds of the fatty acids. Foods of this group are of the highest energy content. Fat-soluble vitamins, particularly vitamins A and E, are naturally present in some unprocessed fats and oils.

Butter and margarine both supply the same amount of food energy—3300 calories/pound. Both are usually excellent sources of vitamin A, the exact amount depending on production and processing variables. Butter varies in vitamin A content seasonally; that produced when the cow is on green feed is much richer in vitamin A than when on dry feed. Fats and oils used in the manufacture of margarine do not possess much vitamin A; however, practically all margarine manufacturers in the United States fortify the products with vitamin A (see Nutritional Enrichment of Food in this chapter).

Vegetable oil is the basic ingredient of most salad dressings. In fact, minimum amounts are required to meet FDA specifications. For example, mayonnaise must contain 65% by weight; French dressing, 35%; salad dressing, 30%. The oil content of commercial dressings may be considerably higher than the minimum requirement (e.g., mayonnaise, 75–80%; French and salad dressing, 35–40%). The content of other nutrients varies depending on the particular formulation involved.

The need for more reliable and comprehensive data on the nutrient composition of our food supply is urgent. Fortunately, the U.S. Department of Agriculture, through its Nutrient Analysis Laboratory, has undertaken a massive effort on this subject. Vastly improved and automated methods of analysis are being developed, and extensive sampling and nutrient analysis of the majority of the foods consumed in this country are currently under way. Within a few years a computerized bank of nutrient content data should be available for easy retrieval and use.

Effects of Processing and Preservation
Practices on Nutritive Values

All of the steps employed in the processing and preservation of foods have an impact on their nutritive value. Some may improve it, but for the most part they have adverse effects. What are the principles involved in controlling the nutritive value? Some examples of good and poor practices will be given.

First, consider the possibilities for improving nutritive value of foods by processing. From the discussion of nutritional inhibitors, we now know this can be done. A few examples will suffice to show how: 1) the nutritive value of egg white is significantly improved by thorough cooking (thereby destroying iron binding by conalbumin and biotin binding by avidin); 2) processing of soybeans into bean curd, soy sauce, and other soy products can inactivate the inhibitors present; and 3) cooking destroys the thiaminase found in certain fish.

The nutrient losses that can occur during the handling, processing, and storing of foods largely involve the vitamins and minerals. A number of factors affect such losses. Probably the greatest loss—and the one least appreciated by most ordinary consumers—is the removal of nutrients by the so-called refining processes used in converting raw agricultural products into food. In addition to these losses, there are those due to physical and chemical action that take place during processing, storage, and distribution.

Cereal Products

The evolution of cereal milling practices over the centuries has led to a significant reduction in the nutritive value of the end products, especially for products made from wheat and rice. As white bread became more and more the preferred product, millers made ever greater efforts to remove the bran and other layers from the wheat and to employ bleaching and other chemical processes in making white flour. As a result, the nutritive value of flour (and of course the products made from it) has progressively declined. "Extraction" is the term commonly used to express the degree of refining. A 100% extraction flour is a product made up of the entire kernel; a 60% extraction flour is a product made up of what is left after 40% of the kernel components have been removed. A 72% extraction flour, which is commonly used in the United States for producing white bread, contains only about a third of the iron, a fourth of the thiamin and niacin, and about a third of the pantothenic acid originally present in the kernel. From these figures, one can see how serious nutrient losses due to refining can be. (See Fig. 42 for a specific study.) The story of milling rice to produce white rice is strikingly similar.

Fortunately for consumers in the United States, industry and government have taken steps to partially correct the problem created by these milling practices used for white flour and rice. The "enrichment" program begun during World War II was sponsored by government and industry and carried out by the latter; it calls for white flour to contain whole-wheat equivalent amounts of thiamin, riboflavin, niacin, and iron. This has been accomplished

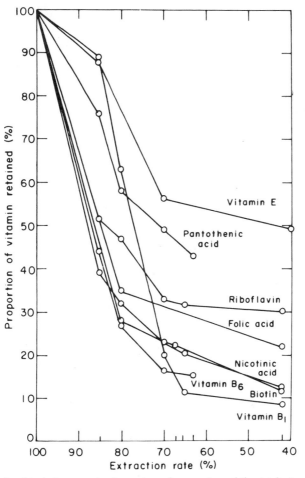

Fig. 42. Relationship between extraction rate and proportion of the total vitamins of wheat retained in flour. (Adapted from Moran, 1959.)

by adding these nutrients to the flour at the mill. Today practically all of the white flour and bread sold at retail in the United States is fortified in this way. To a lesser degree, fortification is used for rice, cornmeal, and pasta products (see Nutritional Enrichment of Food, in this chapter).

Milk

Milk is almost always heat pasteurized to destroy pathogenic microbes before distribution or further processing into dairy products. This process plus handling practices adversely affects the ascorbic acid and thiamin content. In addition, beverage milk was formerly packaged in glass bottles. This type of packaging led to losses in riboflavin if the degree of exposure of the bottled milk to UV light was sufficiently great. The present usage of plastic-coated cartons for milk eliminates this particular problem, because these materials are impervious to the UV light. It should be pointed out here that the losses of thiamin and ascorbic acid in pasteurized milk are not considered very important, nutritionally speaking, because milk is not an important source of ascorbic acid or thiamin in the United States. We usually obtain adequate amounts of these vitamins from other foods (e.g., for thiamin—bread and meat; for ascorbic acid—citrus fruit and tomatoes). However, milk is an important source of riboflavin and care should be and has been taken to avoid appreciable losses from UV light-induced oxidation.

Fresh and Frozen Fruits and Vegetables

Vegetables are an especially good source of several vitamins and minerals, making it important that they be handled so as to minimize losses of these nutrients. Table 21 lists the important factors affecting losses in some of the key vitamins found in food. A few words of explanation are in order. Water solubility is important because water is so frequently used in processing fruits and vegetables (e.g., for washing, conveying, cooking, blanching, and grading). Oxidation is also significant because oxygen is present in these foods as harvested and also because it is so easily incorporated into the food during processing. Heat is important because it is used in blanching and heat sterilization, and it may be a factor if storage of the finished products is carried out at warm temperatures. Finally, UV light is important because the products may be exposed to sunlight or excessive fluorescent light used during raw material handling, sorting, inspection, and grading operations.

Even the relatively short times between harvesting and processing may

TABLE 21

Factors Affecting Stability of Certain Vitamins in Food

Vitamin	Solubility in water	Subject to oxidation	Heat-labile	UV Light-sensitive
Vitamin A	No	Yes	No	Slight
Thiamin	Yes	No	Yes	No
Riboflavin	Yes	No	No	Yes
Niacin	Yes	No	No	No
Ascorbic acid	Yes	Yes	No[a]	Slight
Vitamin D	No	Yes	No	No

[a] If no oxygen is present. However, in the presence of oxygen the heat damage is very extensive.

lead to significant vitamin losses in fruits and vegetables. Table 22 shows the effect of several temperature–time regimes on the ascorbic acid content of green peas and snap beans. These data make it quite clear that both temperature and time are important in controlling losses. Significantly lower losses occur at refrigerator temperatures, but even here approximately half of the ascorbic acid in these vegetables is lost within a week.

Storage losses in vitamin content have also been observed in frozen products. Figure 43 and Table 22 show that, unless frozen peas are kept below −17.8°C (0°F), there is a progressive loss in the ascorbic acid content. Other vegetables would also be expected to show losses during cold storage.

Blanching also affects the vitamin and mineral content of vegetables. This heat process is applied to vegetables before freezing or canning to inactivate the enzymes that cause darkening and off-flavors.

TABLE 22

Effect of Storage on Ascorbic Acid Content of Green Peas and Snap Beans[a]

Vegetable	Temperature °F (°C)	mg/100 gm Storage time (days)		
		0	2	6
Green peas	1.1– 3.3	17	16	10
	7.8– 8.9	20	16	10
	21.1–28.9	17	9	7
Snap beans	1.1– 3.2	18	15	8
	7.8– 8.9	18	11	6
	21.1–28.9	18	7	4

[a]Adapted from Clifcorn (1948).

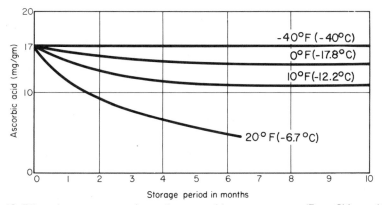

Fig. 43. Effect of storage on ascorbic acid content of frozen green peas. (From Clifcorn, 1948.)

Canned Fruits and Vegetables

Heat sterilization used in the canning process causes significant losses in the vitamin content of foods so preserved. Table 23 summarizes the data on vitamin losses in canned prepared meals (Hellendoorn et al., 1969). These data are the average of six types of food that differed somewhat in their behavior. Nonetheless, the figures show that vitamin losses may vary widely— from 1 to 100%—and that the storage time after canning sometimes leads to additional losses.

Bender (1978) has reviewed the work that has been carried out on vitamin losses during blanching using hot water or steam. Table 24 summarizes the findings for broccoli and brussels sprouts (Hellendoorn et al., 1969). It is clear that blanching causes significant losses, whichever method is used. The results are not entirely consistent, but it would appear that steam blanching is superior in retaining the vitamin.

A number of studies have been made of commercial citrus juice products. The results indicate that ascorbic acid retention ranges from 89 to 100%. Canned tomato juice does not show such excellent retention; its values range from 35 to 90% and average around 67%. Ascorbic acid in tomato juice appears to be especially susceptible to oxidation, particularly while hot. Very careful attention must be paid to avoid oxygen and copper contamination during processing. The use of low temperature during processing and proper steaming to remove oxygen prior to sterilization increases retention of ascorbic acid.

TABLE 23

Average Percentage Losses in Vitamin Content for Six Types of Meals Due to Canning and Subsequent Storage[a]

| Vitamin | Initial value | Percentage loss | | | |
		After canning	1.5 years	Storage 3 years	5 years
Vitamin A	16.5 μg	50	100	—	—
Vitamin E	80 mg	0	0	50	50
Thiamin	9 mg	50	75	75	75
Riboflavin	6 mg	0	0	0	0
Pyridoxine	5 mg	0	0	0	0
Vitamin B_{12}	18 μg	0	0	0	0
Niacin	110 mg	10	20	20	20
Pantothenate	21 mg	25	50	50	50
Folic acid	14 μg	0	0	0	0
Inositol	26 mg	0	0	0	0
Choline	27 mg	0	0	0	0

[a]From Hellendoorn et al. (1969).

TABLE 24

TABLE 24
Percentage Losses of Ascorbic Acid in Two Vegetables during Blanching[a]

Vegetable	Method	Temperature °C (°F)	Time (minutes)	% Loss
Broccoli	Water	100° (212°)	3	30–40
	Water	100° (212°)	4	50
	Steam	—	4	20
	Steam	—	6	40
Brussels sprouts	Water	100° (212°)	3	25
	Water	100° (212°)	5.5	20
	Steam	—	4	15

[a]Adapted from Hellendoorn et al. (1969).

Sterilization is especially destructive of thiamin. Table 25 shows the adverse effect of this treatment on the retention of thiamin in several vegetables. Similar results have been noted with meat, and this fact is especially important because pork is an important source of thiamin in our diet (Farrer, 1955).

Dehydrated and Dry Foods

Dehydration can have adverse effects on nutritive value. Dried fruit is one of the oldest of the dehydrated foods; accordingly, it has received the most attention from nutrition researchers. The data that have been developed show that ascorbic acid is readily lost, both in sun-drying and artificial dehydration of fruits. If sulfur dioxide is used (primarily to prevent darkening), these losses are much lower. (SO_2 is widely used in fruit drying and is also quite common in vegetable dehydration.) Dried vegetables show a pattern similar to fruits. Where sulfur dioxide is used, there are serious losses in thiamin (see Table 26). Both dried fruits and vegetables retain their vitamin values quite well during storage.

TABLE 25
Thiamin Retention of Several Canned Vegetables during Sterilization[a]

Vegetable	Process conditions	% Retention
Asparagus	14 minutes at 120°C (248°F)	66
Whole-kernel corn	30 minutes at 121°C (250°F)	47
Snap beans	20 minutes at 115°C (240°F)	79
Green peas	35 minutes at 115°C (240°F)	64

[a]Adapted from Clifcorn (1948).

TABLE 26
Effect of Sulfur Dioxide on Vitamin Content of Dehydrated Carrots[a]

	Vitamin content (mg/100 gm)	
Vitamin	Control	SO_2-treated
Thiamin	0.41	0.13
Riboflavin	0.37	0.37
Niacin	2.4	2.7
Ascorbic acid	189	351

[a]From Pavcek (1946). Reprinted with permission from *Ind. Eng. Chem.* **38**, 853–856. Copyright 1946, American Chemical Society.

Dried milk and eggs retain their vitamin values very well during spray-drying, providing overheating and scorching are avoided. The latter is particularly damaging to protein quality. There is very little loss of nutrients in these products during storage at room temperature or lower.

The processing of grains to produce dry breakfast cereals must be carefully controlled to avoid damage to the protein quality. Even so, it is difficult to prepare this type of cereal without damage to protein quality.

Combined Treatments

A few studies have been made of overall losses during processing, preservation, and final preparation steps. Table 27 provides data on ascorbic acid losses for fresh, frozen, canned, and dried peas, showing the losses after each processing step and also after the final cooking step. The total loss of ascorbic acid for the cooked fresh peas vs. those after processing, preservation, and cooking were quite similar. The steps involving a heat treatment (blanching, canning, air drying, final cooking, or heating) were obviously the most destructive of ascorbic acid.

TABLE 27
Percentage Loss of Ascorbic Acid in Green Peas Due to Processing, Preservation, and Cooking[a]

Fresh		Frozen		Canned		Air-dried		Freeze-dried	
Process	% Loss	Process	% Loss	Process	% Loss	Process	% Loss	Process	% Loss
—		Blanching	25	Blanching	30	Blanching	25	Blanching	25
—		Freezing	25	Canning	37	Drying	55	Drying	30
—		Thawing	29	—	—	—	—	—	—
Cooking	56	Cooking	61	Heating	64	Cooking	75	Cooking	65

[a]Adapted from Matson (1956).

Butter and Margarine

Loss of vitamin A in margarine during frying has been studied. Figure 44 shows the loss as a function of time of heating. Losses were rapid and extensive under these conditions. One might expect similar results for butter fortified with vitamin A.

Canned Meat

Because thiamin is heat labile, a number of studies have been made on the effect of cooking and canning meat, a major source of the vitamin. Table 28 shows some of the data from one such study. Losses were considerable for all meats, with the highest losses associated with canning and frying.

Nutritional Enrichment of Food

Because of the current confusion resulting from the use of several different terms to describe the addition of pure nutrients to processed foods, we need

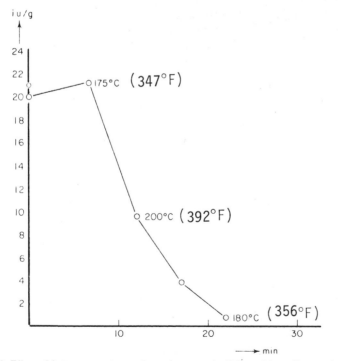
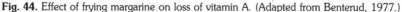

Fig. 44. Effect of frying margarine on loss of vitamin A. (Adapted from Benterud, 1977.)

TABLE 28
Losses of Thiamin in Cooking Meat[a]

Meat type and cooking method		% Loss
Beef:	roast	40–60
	broiled	50
	stewed	50 (up to 70)
	fried (variable conditions)	0–45
	braised	40–45
	canned (85 minutes at 121°C)	80
Pork:	braised	20–30
	roast	30–40
Ham:	baked	50
	fried	50
	broiled	20
	canned	50–60
Chop:	braised	15
Bacon:	fried	80
Mutton:	broiled chop	30–40
	roast leg	40–50
	stewed lamb	50
Poultry:	roast chicken, turkey	30–45
Fish:	fried	40

[a]Data from figures assembled by Farrer (1955).

specific definitions. The term enrichment is used to describe the process of adding pure nutrients to food. It also has legal implications in the United States for certain foods discussed later. Fortification is also used for this same purpose but usually only where the nutrients being added are not normal components of the food or are naturally present only in small amounts. Nutrification is a term coined more recently and is generally used where nutrients are added to certain formulated foods to provide a special contribution to their nutritive value. Restoration, another term that is sometimes used, refers to the addition of those nutrients to food that were lost during refining, processing, and preservation (e.g., the loss of certain vitamins and minerals in the milling of cereal grains). The terms enrichment and fortification are most commonly used in the United States, most of the time interchangeably. We will follow this custom.

Rationale for Enrichment of Food

Concerning the legal aspects, Bender (1978) has proposed a valuable procedure for approving regulated enrichment programs so that they are governed by a certain logic: 1) a demonstration of need for the added nutrient(s) by consumers of the product; 2) selection of a proper vehicle for the added nutrient(s); 3) avoidance of adverse effects of the added nutrient(s) on the palatability of the food; 4) availability of a cost-effective technology for the enrichment process; and 5) effective enforcement of enrichment regulations. The Food and Nutrition Board of the National Academy of Sciences endorses the addition of nutrients to foods when it is in keeping with the following circumstances:

1. The intake of nutrient(s) is below the desirable level in the diets of a significant number of people.

2. The food(s) used to supply the nutrient(s) is likely to be consumed in quantities that will make a significant contribution to the diet of the population in need.

3. The addition of the nutrient(s) is not likely to create an imbalance of essential nutrients.

4. The nutrient(s) added is stable under proper conditions of storage and use.

5. The nutrient(s) is physiologically available from the food.

6. There is reasonable assurance against excessive intake to a level of toxicity.

7. The additional cost is reasonable for the target consumer.

Historically in the United States the logic just outlined has not always been followed, for a variety of reasons—not the least of which have involved politics or vested interests. This has sometimes had an effect on what nutrient(s) are legally required for enrichment and at what level(s).

Regulated Enrichment. Legally mandated enrichment of food is controlled in the United States by regulations promulgated and enforced by the FDA under its "standards of identity" program. These involve only a certain number of our common foods (see Chapter 9 dealing with U.S. laws and regulations). Examples include the addition of vitamin D to beverage milk, vitamin A to margarine, iodine to salt, and thiamin, niacin, riboflavin, and iron to flour, bread, and pasta products. (It is worth noting here that, in the case of enriched flour/bread, meeting these legal requirements does not make up for all of the vitamins lost in flour milling. For example, pantothenic acid, biotin, folic acid, pyridoxine, and α-tocopherol are also lost in the refining process. In addition, wheat contains insignificant amounts of riboflavin and iron, which, however, are required to be added to meet the regulations.)

Voluntary Enrichment. Foods for which there are no FDA standards of identity may be enriched with certain nutrients, provided that the food package label clearly shows them in the list of ingredients. In addition, if nutritional claims are made for the enriched product, then the label must also show the amounts (as percentage of RDAs) of each added nutrient (and other nutrients) in the food as sold to consumers.

Voluntary enrichment of certain foods seems justified in some cases, based on the rationale just given for mandated enrichment. However, a good many of the voluntary enrichment programs seem directed more at "image building" for the particular brand of product than for meeting a demonstrated need for the added nutrients. This seems particularly true for breakfast cereals and for beverages that simulate or resemble orange juice or lemonade.

Nutrification of certain foods is practiced in the United States, particularly for infant formulas, meal-replacement items, and certain special dietary foods. These products usually contain a certain proportion of the RDAs for the particular type of consumers using these products.

Enrichment Technology

Extensive research and development have been carried out by food firms and especially by their suppliers in order to devise effective ways and means of food enrichment. Special efforts have been made to 1) find the most suitable chemical form of the nutrient to use, 2) develop effective carriers for the nutrient(s), 3) develop formulations possessing satisfactory stability for the added nutrients, 4) avoid palatability problems caused by the added nutrients, and 5) develop cost-effective methods of carrying out the enrichment.

The details of the methods used in the commercial enrichment of foods are mostly privileged information—that is, they belong to the firm(s) involved—and thus cannot be meaningfully discussed here. Nonetheless, they are extremely important to an effective and successful enrichment program. In this connection it seems worth noting here that the laws and regulations surrounding the commercial enrichment of food and their enforcement appear to have been effective in protecting consumer interests. (For more information, see National Academy of Sciences, 1975.)

REFERENCES

Beaton, G. H., and Patwardham, V. N. (1976). Physiological and practical considerations of nutrient function and requirements. *In* "Nutrition in Preventive Medicine" (G. H. Beaton and J. M. Bengoa, eds.), p. 445–481. World Health Organ., Geneva.

Bender, A. E. (1978). "Food Processing and Nutrition." Academic Press, New York.

Benterud A. (1977). Vitamin losses during thermal processing. *In* "Physical, Chemical, and Biological Changes in Food Caused by Thermal Processing." (T. Høyem and O. Kvåle, eds). Appl. Sci. Pub.

Canfield, and Liu (1965). The disulfide bonds of egg-white lysozyme. *J. Biol. Chem.* **240**, 1977.

Clifcorn, L. E. (1948). Factors influencing the vitamin content of canned foods. *Adv. Food Res.* **1**, 39–100.

Darby, W. J., Ghalioungui, P., and Grivetti, L. (1977). "Food: The Gift of Osiris," 2 vols. Academic Press, New York.

Drummond, J. C., and Wilbraham, A. (1958). "The Englishman's Food; a History of Five Centuries of English Diet." Cape, London.

Farrer, K. T. H. (1955). Thermal destruction of vitamin B in foods. *Adv. Food Res.* **6**, 257–306.

Food and Nutrition Board, National Academy of Sciences/National Research Council (1980). "Recommended Dietary Allowances," 9th ed. Nat. Acad. Press, Washington, D.C.

Hellendoorn, E. W., *et al.* (1969). Effect of sterilization and three years' storage on the nutritive value of canned prepared meals. *Voeding* **30**, 44–63.

Matson, L. W. (1956). Effects of processing on the vitamin content of foods. *Br. Med. Bull.* **12**, 73–77.

Moran, T. (1959). Nutritional significance of recent work on wheat flour and bread. *Nutr. Abstr. Rev.* **29**, 1–16.

National Academy of Sciences (1975). "Technology of Fortification of Food," Proceedings of a Workshop. Natl. Acad. Sci., Washington, D.C.

Pavcek, P. I. (1946). Nutritive value of dehydrated vegetables and fruit. *Ind. Eng. Chem.* **38**, 853–856.

Peterson, M. S., and Johnson, A. H. (1978). "Encyclopedia of Food Science." Avi, Westport, Connecticut.

Sebrell, W. H., Jr., and Haggerty, J. J. (1967). "Food and Nutrition." Time, New York.

Stewart, K. K., ed. (1979). "Proceedings, Nutrients Analysis Symposium." Assoc. Off. Anal. Chem., Arlington, Virginia.

U. S. Department of Agriculture (1959). "Yearbook of Agriculture—Food." U.S. Gov. Print. Washington, D.C.

Vedder, E. B. (1913). "Beriberi." Wood, New York.

SELECTED READINGS

Adam, W. B., Horner, G., and Stanworth, J. (1942). Changes occurring during the blanching of vegetables. *J. Soc. Chem. Ind., London, Trans. Commun.* **61**, 96–99.

Bender, A. E. (1966). Nutritional effects of food processing. *J. Food Technol.* **1**, 261–289.

Clifcorn, L. E., and Heberlein, D. G. (1944). Thiamin content of vegetables. Effect of commercial canning. *Ind. Eng. Chem.* **36**, 168–171.

Hegsted, D. M., ed. (1976). "Present Knowledge of Nutrition." Nutr. Found., New York.

Lamb, F. C., Pressley, A., and Zuch, T. (1947). Nutritive value of canned foods. Retention of nutrients during commercial production of various canned fruits and vegetables. *Food Res.* **12**, 273–287.

National Academy of Sciences–National Research Council (1980). "Toward Healthful Diets." Natl. Acad. Sci., Washington, D.C.

Rice, E. E., and Beuk, J. F. (1953). Effects of heat on the nutritive value of protein. *Adv. Food Res.* **4**, 233–271.

Souci, S. W., Fachman, W., and Kraut, K. (1981). "Food Composition and Nutrition Tables 1981/82." Wissen Verlagsges, Stuttgart.

Tannenbaum, S. R. (1979). "Nutritional and Safety Aspect of Food Processing." Dekker, New York.

U.S. Department of Agriculture (1966). "Yearbook of Agriculture—Protecting Our Food." U.S. Gov. Print. Off., Washington, D.C.

Watt, B. K., and Merrill, A. L. (1963). Composition of foods. *U.S. Dept. Agric., Agric. Handb.* No. 8, pp. 1–190. (Currently being revised in sections.)

White, P. L., ed. (1974). "Nutrients in Processed Foods. Vol. 1: Vitamins–Minerals; Vol. 2: Proteins; Vol. 3: Fats and Carbohydrates." PSG Publ., Littleton, Massachusetts.

Chapter 7

Shelf Life of Processed Foods and General Principles for Control

Few problems more fully occupy the minds and energies of the food scientist than those of providing an adequate shelf life for processed foods. Understanding the nature of these problems and learning the principles by which they may be solved will be the main focus of this chapter. The more applied aspects of commercial food processing and preservation will be dealt with in Chapter 8.

Quality losses in processed foods may be classified into three types: 1) deterioration of quality, 2) microbial spoilage, and 3) biological contamination. Quality deterioration is caused by physical, chemical, and biochemical reactions within the food and/or by the interaction of environmental chemicals with food components. Spoilage consists of changes in foods brought about by the action of living microorganisms: bacteria, molds, and yeasts. Contamination is caused by pests: insects, rodents, and even birds. These distinctions in the several types of quality losses encountered in food are not always made, even by some food professionals. However, we think it essential to do so because it leads to a better understanding of quality losses as well as an environment for developing principles by which losses may be controlled.

The reactions causing deterioration generally result in a progressive loss of palatability, nutritional value, and/or the functional properties of food. Changes eventually may lead to a level of quality that is unacceptable to consumers. On the other hand, the changes in quality brought by spoilage organisms and pests generally take place much more rapidly and usually are sufficiently objectionable that most consumers would regard the food so affected as unsuitable for consumption.

Before going any further in this discussion, we want to make clear that we are not discussing the safety of such foods. The reasons for this is that food can deteriorate very markedly in sensory quality or can even spoil without creating a health hazard. It is very important to keep this fact clearly in mind as we discuss these quality losses further.

The purpose of this chapter is 1) to examine in some detail quality deterioration, spoilage, and contamination problems in food; 2) to discuss the principles the food scientist applies in preventing or at least minimizing these quality losses; and 3) to provide examples of processing and preservation methods currently in use that maximize the shelf life of processed foods.

Fresh vs. Processed Foods

What we are about to discuss is the deterioration, spoilage, and contamination of *processed* food, not those so-called fresh products, such as apples, bananas, tomatoes, peas, sweet corn, and spinach, found in the produce section of the supermarket. Deterioration and spoilage phenomena in unprocessed fruits and vegetables are somewhat different from their processed counterparts because the former are living entities with actively functional metabolisms and respiration. At the same time we want to make clear that deterioration and spoilage of fresh fruits and vegetables are very important to the food scientist, especially in the selection and handling of raw materials for processing. We will deal with these preprocessing quality losses in Chapter 8, which is concerned with the selection of food raw materials and their transformation into processed products.

Quality Deterioration

Losses in food quality due to deterioration take a variety of forms. Certain losses are caused by the intervention of physical forces, a variety of ordinary chemical reactions, and a considerable number of biochemical (enzyme-catalyzed) reactions. We shall provide examples of each of these processes that are responsible for quality losses in food.

Losses Due to Physical and Physio-chemical Changes

Moisture Loss/Gain. The gain or loss of moisture is a common cause of deterioration in processed foods. For example, look what happens to cheddar cheese when the package is opened and left in a dry place for a few days, even in the home refrigerator. Moisture quickly evaporates from the cheese surface, causing unattractive changes in appearance, cracking, and discoloration. It certainly no longer is an appetizing-looking food.

Moisture pickup also causes losses. In fact, it is one of the most common causes of deterioration in dry or dehydrated foods. A good example is what happens to soda crackers when the package is left open during humid weather. The crackers quickly absorb moisture from the atmosphere and lose their crispness and other desirable textural properties. What a simple cause for a serious loss of palatability!

Aroma Losses. Losses of aroma from foods are fairly common. As an example, remember what freshly ground pepper smells like, or the rich aroma of a freshly opened can of roasted coffee. Both products lose their fine aroma rather quickly when the container in which they are packed is left open, especially in warm weather. Again, here is a rather simple cause for a significant loss of palatability.

Foreign Odor Pickup. Another fairly common cause of quality deterioration, especially during commercial food handling and storage, is foreign odor pickup. A good example would be the absorption of onion odors by butter when these products have been stored together.

Loss of Carbonation. A physio-chemical reaction that causes a serious loss of quality in certain beverages is loss of carbonation. For example, have you ever tasted a carbonated soft drink from which all of the carbon dioxide has escaped? Terrible! An even better example is the "flat" taste of beer that has lost its carbonation. It would hard to imagine a less appetizing beverage!

Structural Breakdown. The breakdown of structure is still another quality loss caused by a physio-chemical reaction, although, it is not as common as others just mentioned. For example, certain kinds of sauces used in making prepared entrées (e.g., chicken à la king) break down if frozen and thawed, resulting in the separation of components. It is an unsightly mess that most consumers would throw out (Fig. 24).

Crystallization. Yet another structural change that causes a loss in quality of certain foods is crystallization. For example, honey undergoes a crystallization of the sugar present if it is kept cold for too long a time. What was once an attractively colored transparent syrup turns into an opaque half-solidified mass that not only looks unappetizing but is very difficult to use.

There are other physical and physio-chemical changes in food that cause sensory and functional quality losses. However, these examples should suffice to give the student a clear idea of the kind of deterioration we are addressing here.

Chemical and Biochemical Changes

Lipid (Fat) Deterioration. A very serious problem in the preservation of fatty foods is that whether the lipid is in pure form (e.g., salad oil) or is a component of a food (e.g., in nut meats), lipids are readily oxidized, with the result that the food develops a strong off-odor and off-flavor.

Lipid oxidation may be caused by 1) an enzyme-catalyzed (biochemical) or 2) chemical peroxidation reaction. The latter has been somewhat better characterized and its significance in the quality loss in fatty foods is better understood. Peroxidation of lipids is the major cause of rancid flavors in fatty foods. For example, potato chips rather quickly turn rancid from this reaction if held at warm temperatures, and especially if exposed to oxygen (air) and UV light. The objectionable odor and flavor often show up within a few days after opening the package, with the result that the chips are then frequently discarded by consumers. (Some people, however, seem not to object to the odor and flavor of rancid potato chips. It is speculated that these people have experienced the off-flavor over such a long period of time that they have gotten used to it! Such are the vagaries of predicting the acceptability of food.) Another example of peroxidation rancidity may be found in shelled nut meats, especially peanuts, walnuts, and pecans. If the can or jar containing the meats is left open, if the weather is warm, and especially if the nuts are exposed to UV light, a very objectionable odor and flavor develop. Most consumers would discard such nuts as inedible.

An enzyme-catalyzed lipid oxidation occurs in full-fat soy flour. In this case the lipid fraction of the flour is rapidly oxidized by atmospheric oxygen but this time is catalyzed by the enzyme lipoxidase, which is naturally present in soybeans. The resulting reaction causes a rancid odor; most food processors (the principal users of the flour) would discard the flour for formulating purposes because the off-odor will usually carry through into the finished product.

Another type of enzyme-catalyzed lipid reaction leads to a hydrolysis of the fat. This reaction is known to occur in raw milk and especially in butter made from unpasteurized cream. One of the fatty acids (butyric) split off has a strong odor that is generally considered very objectionable. This reaction is catalyzed by the enzyme lipase.

There are other lipid reactions causing off-flavors in foods that have not been well characterized. One of special interest is that found in certain seed oils such as soybean oil. Under certain conditions, these reactions give rise to "fishy" and "painty" off-flavors.

Browning Reactions. There are two types of browning reactions: enzymatic and nonenzymatic (e.g., the Maillard reaction). The most common enzymatic browning reaction occurs in fruits and vegetables and involves the oxidation of the colorless polyphenols by oxygen (from air). It is catalyzed by the enzyme polyphenol oxidase. The reaction causes discoloration in certain freshly cut fruits (e.g., bananas) and vegetables (potatoes). An example of the second type involves a reaction between a reducing sugar (e.g., glucose) and an amine or amino group of another food component. This reaction causes a brown to black discoloration in a number of foods, including dried egg white and dry milk products.

Protein Degradation. Under certain conditions, the soluble proteins of beer can destabilize and precipitate, causing an unsightly cloudiness and/or sediment to form in the beverage. Then there is the enzyme-catalyzed excessive breakdown of proteins that sometimes occurs during the aging of certain cheeses. If carried too far, it results in a bitter taste, presumably due to the formation of bitter-tasting amino acids and/or small peptides.

Starch Degradation. Many starches used as thickening agents in foods lose their viscosity due to chemical hydrolysis. The food products in which they were used become thin or even watery. This is especially true if the food is acid in character, such as certain types of baby food. In bread, especially white bread, starch rapidly undergoes retrogradation, which causes the bread to lose its characteristic elasticity and develop a dry mouth-feel, often referred to as "stale." Most consumers regard such bread as inferior in quality and refuse to eat it. This is why "day-old" bread is sold at a substantial discount in the market. (Incidentally, sourdough French bread, usually made without shortening, stales even faster than white bread for reasons not fully understood.)

Pectin Changes. Serious quality losses in a variety of fruit and vegetable products can result from pectin changes. One such reaction, catalyzed by the action of the enzymes, pectinesterase and polygalacturonase on pectins in the cell wall, leads to a drastic lowering of viscosity in freshly prepared but unheated tomato juice. If the juice is made into catsup, the resulting product is thin and watery, which is quite unacceptable to most consumers. Another degradation results when the pectin in certain fruit juices destabilizes, leading to clouding and sediment formation. This can be a serious problem for juices that are expected to be transparent (e.g., apple juice).

Pigment Degradation. The common plant pigments—chlorophyll, carotenoids, and anthocyanins—are subject to serious deteriorative change from a variety of reactions. We will cite two examples. Chlorophyll, the green pigment, and carotenoids, the yellow-orange-red pigment, both undergo enzyme-catalyzed reactions that cause color fading, and sometimes a complete loss of color. These reactions cause serious quality problems in canned or frozen green vegetables, such as canned peas, and in red-fleshed fish, such as frozen or canned salmon. There are also a number of nonenzymatic reactions in food that involve these pigments.

A cation-exchange reaction of chlorophyll (involving the replacement of magnesium by hydrogen in the molecule) causes an off-color in canned green vegetables. An isomerization reaction (the transformation of one isomer of the compound into another) in carotenoids causes the yellow-orange pigment of pineapple to turn yellow. A metal-complexing reaction (formation of a tin–pigment complex) involving the anthocyanin pigment in canned pears causes the formation of a pink off-color, considered abnormal by most consumers. Many would discard the discolored fruit.

Aroma and Flavor Degradation. Many foods can deteriorate in quality because of aroma or flavor degradation. Unfortunately, most aromas and flavors are very complex mixtures of chemicals, many of which have yet to be fully characterized—making it difficult to study their deteriorative reactions. Thus we have little or no information about the specific reactions that cause the off-flavors. We do know that some of these reactions are enzyme-catalyzed, whereas others are not. We also know how to avoid some but not all of the losses, although much more research is needed in this area. Fortunately it is a very active field of research at writing.

Food Spoilage and Contamination

The reader will recall that food spoilage and contamination have been defined as those adverse changes in quality caused by the action of microorganisms and by insect, rodent, and bird pests. (Viral agents do not grow in food and therefore are not responsible for food spoilage. However, as discussed in Chapter 4, they can cause health hazards in food.) The quality changes in food produced by the microbes and pests vary widely, including drastic changes in appearance, color, odor, flavor, and texture. Since we cannot cover all of the types of spoilage and contamination problems known to occur in food, we will restrict ourselves here to some of the more common and serious types (see Ayers *et al.*, 1979).

Microbial Spoilage

Bacteria. Bacteria are a very common cause of spoilage in food and many species have been implicated. They produce many changes in food, the characteristics of which cover almost the entire spectrum of spoilage types: off-colors, off-odors, off-flavors, textural defects, and gas. The processed foods in which spoilage has been observed in this country include: 1) chilled foods— unprocessed red meat, poultry, and fish; milk, sweet cream, sour cream, buttermilk, yogurt, and cottage cheese; 2) canned foods—tomato paste and sweet corn; and 3) fermented foods—olives, pickles, wine, and beer. Spoilage in dehydrated and frozen foods is uncommon, although if these items are mishandled they too can spoil rather readily from microbial action.

Undesirable color changes in foods brought about by spoilage bacteria vary tremendously, from the normal green, yellow, orange, red, and so on, to almost any other color imaginable, including black. Similarly, a wide range of off-odors and off-flavors result from bacterial action, including some that are highly objectionable, such as sour, "rotten egg" (hydrogen sulfide), putrid, moldy, cheesy; the list is almost endless.

One of the common by-products of bacterial spoilage is gas. The reader may recall seeing bulging cans of food; this is sometimes caused by gas produced by spoilage bacteria (but also from corrosion of the can). The gas and/or the food itself may or may not smell. In any case, consumers are very suspicious of gas in canned foods (where it is not a normal component), and they usually discard such food. (This is a prudent move because the gas could have been produced by a pathogen.)

Yeasts. Yeast spoilage is somewhat less common than that caused by bacteria, but most consumers have seen foods spoiled by yeasts. The characteristics of this type of spoilage vary considerably, but not as much as for bacterial spoilage. Yeast spoilage is not generally regarded as being as objectionable to consumers as bacterial spoilage. Gas production is a common symptom, as are yeasty and alcoholic off-odors and film formation. Yeast spoilage is commonly associated with foods high in sugar and acid, such as dried fruits, jams and jellies, some soft drinks, certain types of chocolate candies, sweet-ened condensed milk, and fermented foods such as pickles and olives.

Molds. Mold spoilage is more common than that caused by yeast but less so than bacterial spoilage. Most consumers are familiar with moldy food; the most common occurrence is among the intermediate-moisture foods such as bread, jams and jellies, cake, cured meats, certain types of cheese, and other fermented dairy products. The most common changes caused by molds are unsightly appearance, off-color, and off-flavor. Most consumers reject moldy foods as inedible, although some have been known to scrape off mold from cheese and eat it—a questionable practice considering the possibility of mycotoxins (see Chap. 4).

Pest Contamination

Insects. Insects frequently cause food losses, damage, and contamination of certain ingredients and certain processed foods. They are most commonly a problem in the dry ingredients used in formulated foods, but on occasion they are a problem in the finished product. For example, weevils are a common insect pest in cereal grains and products made from them, such as wheat flours. These insects consume food and leave behind debris and their excreta. Many are destroyed during processing, but their body fragments and excreta remain in the finished food. A similar story can be told about other insects that may infest dried fruit, nuts, spices, and other ingredients. Foods contaminated with insects, and their debris and excreta are legally considered to be contaminated and therefore inedible. When such pests are found, the food is usually discarded by consumers (users). However, it is sometimes difficult to detect their presence in food by ordinary sensory examination. Government regulations in this country regard insect infestation as evidence of poor sanitary practices in the food-processing plant and of possible contamination of the food with pathogens. Foods so contaminated are therefore subject to seizure and destruction by regulatory agencies. Contamination is usually determined by regulatory agencies, using microscopic examination and/or special chemical tests.

Rodents and Birds. Rodents and birds can cause serious food losses and damage and contamination of food ingredients and finished products. Rodents consume food, shed debris, and leave excreta on or near food. Birds (especially pigeons and sparrows) cause similar problems, although they frequently find it more difficult to gain access to the processing plant. Rodent and bird contamination is not always as easy to detect as that caused by insects. Nevertheless, most consumers (users) consider food that they find to be so contaminated also to be inedible. Government regulations usually stipulate that rodent and bird contamination of food is evidence of unsanitary conditions in the processing plant and of possible contamination of the food with pathogenic microorganisms. Such products are considered adulterated and are thus subject to seizure and destruction (see Chapter 9).

General Principles for Control of Shelf Life

Introduction

A variety of measures have been developed over the years for controlling deterioration in and spoilage of commercially processed food products (see

Heid and Joslyn, 1967). Those used to control deterioration generally only slow down loss of quality, whereas many of those used to control spoilage prevent these quality losses permanently. We will now examine the principles involved in quality-preservation methods.

Physical and Physicochemical Quality Deterioration

Moisture Loss/Gain. The loss or gain of moisture can be rather simply controlled by proper packaging methods. Preventing moisture exchange in foods is best controlled by packaging the foods in metal cans and glass bottles and jars. Flexible packages made up of laminated layers of paper and/or polymer film formed into a pouch and sealed at the edges usually provide only fair protection against moisture gain or loss. However, if aluminum foil of proper thickness is used in the lamination, much better protection is achieved. (For some uses of foil-laminated bags, see Fig. 45.)

Aroma Loss and Pickup of Foreign Odors. This problem can be rather easily controlled, again by packaging in vapor-proof containers such as metal, glass, or film laminates containing aluminum foil of requisite thickness.

Chemical and Biochemical Quality Deterioration

Biochemical. These reactions are commonly controlled commercially by the heat inactivation of the enzyme(s) involved. This method is in common use in the canning and freezing industry and to some extent by the other industry segments. However, in some cases it cannot be used because of unacceptable changes in the quality of the food caused by the heat treatment required. For example, in the production of frozen peach slices, such heat treatment would completely change their appearance and flavor. In such cases chemical treatment can be used to suppress enzyme activity. Sulfur dioxide or ascorbic acid plus citric acid are commonly used as enzyme inhibitors and perform satisfactorily in many products.

Lipid Oxidation. Deterioration from lipid oxidation is rather easily controlled by 1) removal of the oxygen and gas-tight packaging (i.e., vacuum packing or replacing the air in the package with an inert gas, such as nitrogen); or 2) the incorporation of a suitable antioxidant into the food. A typical example of the first method is the vacuum packing of nut meats in metal cans or glass jars, a common commercial practice. A less effective but cheaper package, frequently used for potato chips and other snack foods, is a polymer film or

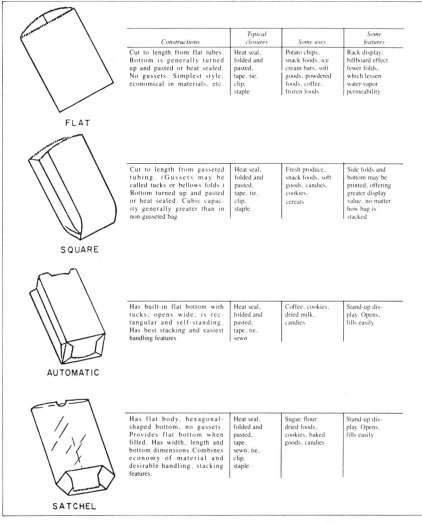

Fig. 45. Standard bag styles. (Courtesy The Aluminum Association.)

glassine bag in which the air has been displaced by nitrogen gas before sealing. Many formulated foods that are subject to oxidative rancidity, such as soup mixes, are protected by incorporating an antioxidant directly into the food. Figure 46 shows the benefit to shelf life of lard afforded by the use of butylated hydroxyanisole (BHA), a synthetic antioxidant. BHA, BHT (butylated hydroxytoluene), and propyl gallate (PG) are three synthetics that are effective and are widely used for this purpose. α-Tocopherol (vitamin E),

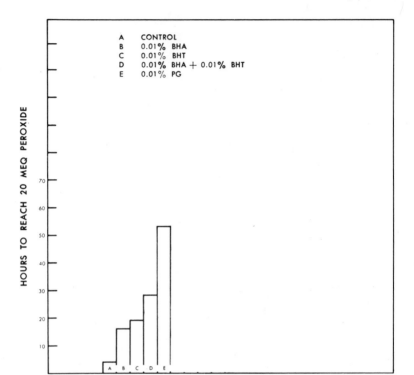

Fig. 46. Effect of BHA on rancidification of lard. Bar A, control; B, 0.01% BHA; C, 0.01% BHT; D, 0.01% BHA plus 0.01% BHT; E, 0.01% PG. See text for abbreviations. (From Stuckey, 1968. Reprinted with permission from *Handbooks of Food Additives.* Copyright The Chemical Rubber Co., CRC Press, Inc.)

although generally a less effective antioxidant than the synthetics, is also sometimes used for this purpose. All are approved for this use by government regulations.

Water Activity. Water activity (A_w) exerts an effect on the development of oxidative rancidity in certain dehydrated foods. Figure 47 shows the effect of A_w on the rancidification of whole-milk powder, where an intermediate value was most effective. Water activity has not been much used commercially for this purpose, but it appears to offer much promise for certain applications in the future.

Low storage temperatures slow down these oxidative reactions, but even freezing temperatures do not stop them completely. Accordingly, refrigeration has not proved to be a primary method for rancidity control.

Maillard Browning. This deteriorative process may be controlled either by removing one of the reactants or by using a chemical inhibitor. In dehydrated

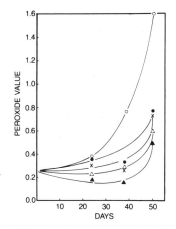

Fig. 47. Effect of A_w on rancidification of spray-dried whole milk at 37°C (98.6°F). ○ = 0.00; ● = 0.42; X = 0.75; Δ = 0.18; ▲ = 0.53. (From Loncin *et al.*, 1968.)

egg products this reaction is controlled by removing the glucose present prior to drying (by fermentation or an enzyme treatment). In dehydrated fruits, such as dried apricots, it is usually controlled by the use of sulfur dioxide or sulfite salts, which combines with the reducing sugars present and prevents reaction with the amine group. Refrigeration and optimal A_w are also useful in controlling these reactions, although neither has proved sufficiently effective for general commercial application. Figure 48 shows the effect of storage temperature on the browning reaction in dried egg white; Fig. 49 shows the effect of A_w on the browning reaction in dehydrated pea soup mix. These results show that A_w control and refrigeration would favorably affect the shelf life of these products.

Starch Retrogradation. The staling of bread caused by the retrogradation of starch has proved extremely difficult to control. For this reason, fresh bread is delivered daily by the baker to the supermarket. Interestingly enough, in contrast to most other deterioration reactions, starch retrogradation is accelerated by chilling. However, it is partially reversed by reheating the bread, a practice familiar to many homemakers.

Reactions Affecting Color. Deteriorative color changes in certain canned fruits and vegetables have been somewhat difficult to control. An effective method for the control of the tin–anthocyanin complex formation mentioned previously (see p. 181) is to prevent the formation of tin ions (from the inner can surface) by using of a suitable polymer coating. Other color problems have not yielded to such simple answers. They present interesting and challenging problems for research workers.

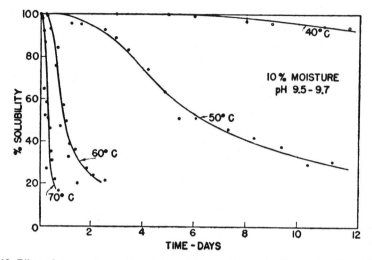

Fig. 48. Effect of storage temperature on browning reaction in dried egg white (From Stewart and Kline, 1948. Reprinted with permission from Ind. Eng. Chem. **40,** 916–919. Copyright 1968, American Chemical Society.)

Protein Degradation. Clouding and sediment problems due to protein degradation in beer (and certain other beverages and juices) can be prevented by treatment with proteolytic enzymes. For example, ficin is used in beer to hydrolyze the protein, which solves the problem. The excessive digestion of protein in certain cheeses mentioned previously (see p. 180) can be prevented by the proper selection of enzymes and careful control over the aging conditions.

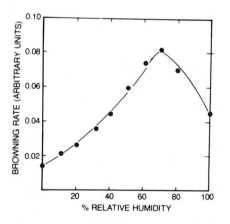

Fig. 49. Effect of A_w on browning of dehydrated pea soup mix at 54°C. (From Labuza *et al.* 1970. Reprinted from *Food Technol.* **24**: 543–550. Copyright © by Institute of Food Technologists.)

Pectin Degradation. Clouding and sediment problems due to pectin degradation may be controlled by the judicious use of pectinolytic enzymes. These enzymes hydrolyze the pectin molecule into fragments that are soluble and that do not precipitate on storage.

Supplementary comments about the control of deteriorative reactions seem to be in order here. Refrigeration has been mentioned as being of supplementary value in controlling these reactions (except for starch retrogradation). With dehydrated foods, cool storage has proved valuable and is used to some extent for storing certain ingredients, such as dehydrated whole egg and yolk and dehydrated chicken. Very low freezing temperatures have proved very valuable in controlling deteriorative changes in many frozen foods (see Fig. 50).

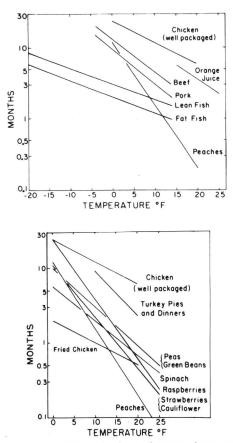

Fig. 50. The shelf life of a variety of frozen foods as a function of storage temperature and time (based on retaining palatability). (Adapted from Shepherd, 1965.)

Control of Microbial Spoilage

Food spoilage is controlled by use of several principles, some similar to those used for the control of deterioration, and some quite different. Let us examine some of the more important ones: 1) heat destruction of spoilage microorganisms, 2) refrigeration to retard or prevent their growth, 3) A_w control, 4) chemical inhibition or destruction of these microorganisms, and 5) the total removal of spoilage organisms from the food.

Heat Destruction

As we have discussed in Chapter 4, heat treatment can be used to destroy all of the microorganisms present (i.e., sterilization) or most of them (i.e., pasteurization). (In the dairy industry, pasteurization has a somewhat different meaning. In that industry it means the destruction of all pathogens but not necessarily all spoilage organisms.) Because we have already discussed the principles used in heat treatment (see p. 96), we need not repeat them here. Figure 51 shows the time–temperature regimes necessary for the complete destruction of a very resistant bacteria that causes "flat sour" spoilage in canned sweet corn. It is interesting to note here that this organism is more heat resistant than are spores of the most feared pathogen, *Clostridium botulinum* (see p. 91–92). This important and fortunate fact means that, for canned sweet corn at least, if we destroy all of these spoilage organisms we need not worry about the safety of the product because all *Cl. botulinum* cells and spores will have been destroyed as well (see Stumbo, 1973).

Some current sterilization and pasteurization practices will be discussed in Chapter 8.

Refrigeration

Chill temperatures greatly reduce the growth of all microbes, as we have already noted in Chapter 4. This permits us to increase substantially the time many foods can be kept without spoilage. This is especially useful for those processed foods that will not withstand the high temperatures required for sterilization: fluid milk, cream, sour cream, buttermilk, yogurt, cottage cheese, red meat, chicken, turkey, fish, shellfish, and certain cured meats. Although chill temperatures provide only limited shelf life for these products, it is sufficiently long (a few days to a few weeks) and the cost is sufficiently reasonable, that the consumer can get these highly acceptable foods the year around. Temperatures at or near the freezing point (30° to −35°F; −1.1° to 1.7°C) of the food are best for this purpose.

Freezing temperatures are extremely valuable for preventing spoilage. However, just freezing the food is not sufficient because some microbes grow

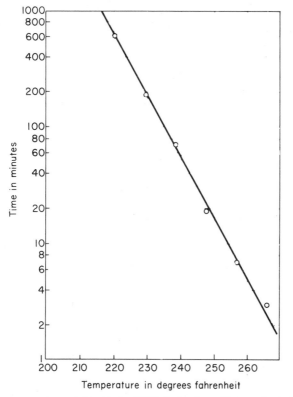

Fig. 51. Time–temperature relationship for 100% destruction of flat sour bacteria in sweet corn. (From Bigelow and Esty, 1920; Copyright by the University of Chicago Press.)

at temperatures as low as $+15°F$ ($-9.4°C$). Accordingly, to prevent the spoilage of frozen foods the temperature must be below that temperature. To accomplish this as well as to minimize deterioration (discussed previously), frozen foods are usually rapidly frozen, and then stored at temperatures of $0°F$ ($-17.8°C$) or below (*see* Van Arsdel *et al.*, 1969).

Water Activity (A_w) Control

Minimum A_w requirements for the growth of pathogenic microbes were discussed in Chapter 4. The same principles apply to the control of spoilage organisms as for pathogens. Generally speaking, spoilage bacteria have high minimum A_w requirements, ranging about 0.95 to 0.99 (almost pure water). Spoilage yeasts have a lower minimum requirement than bacteria, generally near 0.90, but some grow at A_w as low as 0.62. Molds have minima similar to yeasts, ranging from 0.90 down to 0.60. As was also mentioned in Chapter

4, the A_w requirement for microbes is affected by other factors such as the content of sugar, salt, and curing agents, and by pH and temperature. Accordingly, when using A_w to control spoilage, these other variables must be taken into account.

Water activity is usually controlled in food by removing water (concentration or dehydration), by limiting the amount of added water (e.g., in formulation), and by using A_w-depressing ingredients, such as sugar or salt. Much more research must be carried out before we know more precisely how to use A_w effectively in preventing spoilage. The many variables affecting its efficacy make it a complicated task to define the minimum requirements for the many foods where its use looks promising. Research in this field is very active and the future for A_w control in preventing spoilage looks bright (see Troller and Christian, 1978).

Chemical Preservation

The use of chemicals is an important means of inhibiting or preventing microbial spoilage of food. Because of the legal problems involved in establishing safety in use, only a few chemicals that are both effective and safe can legally be used. Chemical preservatives that meet these criteria and are in use in this country are shown in Table 29. Benzoic acid and its derivatives are successfully being used in margarine, soft drinks, salad dressings, baked goods, and certain food specialties. These compounds are very effective in preventing spoilage in these foods and have been shown to be safe in the amounts premitted. Sorbic acid and its salts have also found wide application. Those compounds are particularly effective against spoilage by yeasts and molds, in such products as high-moisture dried fruits, and are safe at permitted levels. Salts of propionic acid are very commonly used in certain cheese products and are safe. (Interestingly enough, propionic acid is a metabolic product of one of the bacteria involved in the production of Swiss cheese; thus this product is not naturally subject to moldiness.) Sulfur dioxide (SO_2) and the sulfite salts are fairly commonly used in soft drinks, fruit juices, and wine, where they are effective in preventing microbial off-flavors. Some question has been raised about their safety, especially because they inactivate the vitamin thiamin. The FDA has set strict limits on their use in foods. For example, SO_2 is not permitted in foods that are considered a dietary source of thiamin.

Physical Methods

Sterile Filtration. The process of sterile filtration can be used for certain liquid products. It was perfected only recently and quickly found application in beer

TABLE 29

Chemical Preservatives Legally Permitted in the United States for Controlling Food Spoilage[a]

Types of food products	Benzoic acid and sodium benzoate	Ethyl and propyl benzoates (parabens)	Sorbates	Propionates	Sulfites	Acetates, diacetates	Epoxides Ethylene oxide	Epoxides Propylene oxide
Beverages								
Soft drinks	+	+	+	−	+	−	−	−
Fruit juices	+	+	+	−	+	−	−	−
Wines and beer	−	+	+	−	+	−	−	−
Purees and concentrates	+	+	+	−	+	−	−	−
Cheese and cheese products	−	−	+	+	−	−	−	−
Margarine	+	−	+	−	−	−	−	−
Baked goods								
Yeast-leavened	−	−	−	+	−	+	−	−
Chemically leavened	−	+	+	+	−	+	−	−
Pie crust and pastries	−	+	+	+	−	+	−	−
Pie fillings	+	+	+	+	−	−	−	−
Processed meat and fish								
Cured meat	−	−	+	−	−	−	−	−
Preserved fish	−	−	+	−	−	−	−	−
Specialties								
Salads, salad dressing	+	+	+	−	−	+	−	−
Dried fruits & vegetables	−	+	+	−	+	−	−	−
Pickles, relishes, olives, and sauerkraut	+	+	+	−	+	+	−	−
Spices	−	−	−	−	−	−	+	+
Nut meats	−	−	−	−	−	−	+	+

[a]Compiled from Code of Federal Regulations (1979).

and wine making. The very small pore size of the filter effectively removes all microorganisms from these beverages, thus eliminating spoilage problems. The product emerging from the filter must be placed directly into containers and sealed aseptically so as to prevent microbial recontamination. This process will probably find many additional applications in liquid foods where its use is practical.

Special Packaging Methods. Food spoilage can be delayed and perhaps prevented by modern vacuum or inert gas packing methods, which can effectively remove all oxygen from food packaged in metal cans and glass jars and bottles. If only strictly aerobic microbes cause spoilage in a given food, this process will prevent spoilage. Unfortunately, there are few such situations in food preservation because facultative anaerobic spoilage organisms, which are able to adapt to anaerobic conditions, are such common contaminants of the environment. However, anaerobic packaging has found a place in extending the shelf life of chilled red meats, especially in the handling of the prime cuts of beef from the meat packer to the retail market. In addition, the use of carbon dioxide in the anaerobic package has been found to provide additional shelf life due to the inhibiting effect of CO_2 on many of the common spoilage bacteria found in meat (see p. 101). (Controlled atmospheres have long been used in the storage of unprocessed fruits such as apples and pears.)

Sanitation

In food-processing operations sanitation is the primary means for minimizing the initial microbial load in food. This can be of considerable value in extending the storage life of most foods, especially chilled items. The reason for this is that the rate of spoilage by microbes is greatly influenced by the numbers found in the product at the beginning of a holding period. This topic will be discussed further in Chapter 8 (see also Troller, 1982).

Pest Control in Food-Processing Plants

The control of pests in and around the processing plant is difficult, time-consuming, and a continuing task. The main techniques used are 1) sanitation, 2) chemical treatment, and 3) certain packaging methods. We will examine each of these briefly.

Sanitation not only involves keeping the plant premises and the equipment clean and sanitary, it also means providing suitable rodent- and bird-proof areas for the storage of raw materials and other ingredients. In some cases this requires the use of well-designed storage areas (e.g., tightly and finely screened) to prevent the pests from gaining access to the ingredients. This method is especially useful for the protection of ingredients that must be kept on hand more or less continuously, such as spices, condiments, salt, flour, and sugar.

Pesticides are very commonly used for the control of insects and rodents. They have their limitations, however, largely because of safety and recontamination problems. A variety of fumigant chemicals, some of them highly toxic, are used for the control of pests in food plants. For cereal products, nuts, dried fruit, spices, and condiments, ethylene oxide, propylene oxide, ethyl formate, and methyl bromide are widely used; however, each compound is used for a specific product(s). Because all of these compounds are toxic and ethylene oxide is also explosive, great care must be exercised in their use. Usually specially trained pest-control operators are employed to carry out the fumigation. They are generally licensed by a regulatory agency and thus operate under rather strict supervision.

Rodent control is also accomplished by trapping or by the use of special baits. The safest of the latter types contain anticoagulant chemicals. They function by causing lethal hemorrhaging in the rodent when consumed over a period of time. As used, they do not present a health hazard to humans because the latter do not come into direct contact with the bait.

Control of birds in the processing plant is generally accomplished by careful screening of openings through which birds could enter.

Refrigeration has a limited application in pest-control programs. Common pests generally avoid cold environments, especially temperatures below freezing. Also, most are destroyed by exposure to temperatures below 20°F ($-6.9°C$) for a period of time.

Food Irradiation

Historical Background

X-Rays were discovered in 1895 and radioactivity in 1896. X-Rays and gamma-rays represent forms of electromagnetic radiation. Radioactivity also produces two other forms, α and β-rays, that are particulate in nature. All four forms are commonly referred to as ionizing radiation.

Soon after the discovery of ionizing radiation, scientists began to learn of their profound effects on biological systems. In particular, they discovered that ionizing radiation is especially lethal to microorganisms. This finding early suggested the possibility of using irradiation to preserve food. Results of a number of these studies are to be found in the scientific literature of the early part of the twentieth century. In 1930 a French patent was issued covering the use of X-rays for preserving food in sealed containers (Wüst, 1930).

Progress in the development of food irradiation was impeded in the early days by a lack of large sources of radiation. However, in the 1930s and 1940s this situation changed dramatically with the development of large X-ray and electron beam generators. The newly developed van der Graaf generator permitted workers at the Massachusetts Institute of Technology in 1943 to irradiate hamburgers. They discovered that this meat could be sterilized by high doses of radiation from the generator (Proctor et al., 1943).

Considerable research and development in the field of food irradiation, sponsored chiefly by the U.S. government, was initiated shortly after World War II. Since that time tens of millions of dollars and hundreds of man-years of effort have gone into this development. These studies have provided a great deal of information about the beneficial effects of irradiating food but also have uncovered many problems and limitations of this new technology. By 1981, after more than 30 years of study, we may be ready to witness the commercial irradiation of food in the United States. However, the FDA must formulate and issue regulations permitting its legal use. This has not yet happened and we will be discussing some of the reasons later.

The Irradiation Process

Ionizing radiation produces its effects by breaking chemical bonds in the target compounds. This leads, directly and indirectly, to chemical reactions that can be numerous and complex in biological systems. Some of these reactions in food produce beneficial effects, others detrimental ones. For example, microorganisms are rather readily destroyed by ionizing radiation, as are many parasites. Insects are also very vulnerable to ionizing radiation. In higher plants metabolism is altered by irradiation, giving rise to beneficial as well as detrimental effects depending on the plant and the dosage employed. In food, many reactions detrimental to the sensory properties and nutritional values are initiated by high-dose irradiation, but a few of the reactions can actually improve the sensory quality of certain foods.

The problem in making use of ionizing radiation in food processing is to find ways to maximize the beneficial effects while minimizing or preventing the detrimental ones. We will be addressing this problem next.

Dose Considerations

The amount of radiation received by an irradiated object is called the "dose." The basic unit is the Gray (Gy); one Gray equals one joule (unit of energy) per kilogram of irradiated material. In the irradiation of food, workers commonly refer to the dose as being "high" or "low." A high dose is more than 10,000 Gray; a low dose is less than 10,000.

Research workers in the field have determined the approximate dose levels required to accomplish the beneficial effects. They are as follows (see Urbain, 1978, for details).

Effect desired	Dose range (Gy)
Sterilization	24,000 to 43,000
Pasteurization	1,000 to 5,000
Destruction of pathogenic microorganisms	4,000 to 10,000
Destruction of pathogenic parasites	200 to 10,000
Destruction of insect pests	150 to 750
Delay in sprouting in tubers and bulbs	50 to 150
Delay of senescence in fruit	350 to 1,250

These figures make evident the fact that the dosage requirements for beneficial effects vary widely, depending on the desired effect as well as the particular food being treated. Sterilization obviously requires the highest doses. Research has also shown that it is at these high levels that serious side effects begin to show up, especially in foods of animal origin. Off-flavors, off-colors, and textural defects are some of the side effects caused by high-dose irradiation of such products as meat, poultry, eggs, and dairy products. Sometimes the side effects can be minimized; sometimes they cannot.

Potential Applications—Low Dose Levels

Properly packaged meats and fish, if irradiated with low doses and kept under refrigeration, have a substantially longer shelf life than similar products that are unirradiated. Food-borne pathogens (e.g., *Salmonellae*) can be substantially reduced or eliminated by low-dose irradiation. In addition, parasites (e.g., *Trichina*) are readily destroyed by low-level irradiation.

Unprocessed fruits and vegetables irradiated at low dose levels can be rendered free of insects; a substantial reduction in spoilage microorganisms can also be accomplished. In addition, senescence (overripening) in fruit and sprouting in tubers ane bulbs can be delayed substantially. These effects provide for a significant increase in the shelf life of many produce items.

Cereal products can be freed of insect pests by low-dose irradiation.

In all of these low-dose applications there are substantially no adverse side effects from irradiation.

Potential Applications—High Dose Levels

Meat and fish hermetically sealed in metal cans can be sterilized by high-dose irradiation. These products keep indefinitely at room temperatures and are superior in sensory quality to similar heat-sterilized products. However, to obtain the optimal quality it is necessary to irradiate these products while frozen and certain additives must be used to protect texture. Naturally these additional requirements increase production costs.

Spices and condiments can be sterilized by high-dose irradiation. This greatly enhances their value for use in producing formulated foods because of the high microbial loads usually found in commercial spices and condiments.

Hospital diets for patients highly susceptible to infections can be sterilized by high-dose irradiation. Although not generally as acceptable as regular hospital food, these irradiated diets seem to be acceptable to such patients and, of course, do protect them from infections due to food-borne pathogens.

Regulatory Problems and Future Outlook

Regulatory officials have been very cautious and deliberate in considering whether to approve the commercial production of irradiated foods in the United States. The FDA early described food irradiation as a food additive rather then simply as a process. This makes it necessary for food firms that plan to employ irradiation to petition the FDA for permission. The petition must include the results of animal-feeding studies to demonstrate the safety of the foods that have been irradiated.

The FDA has three major concerns about irradiated foods: 1) possible creation of toxic compounds in the food, 2) presence of microbial health hazards in irradiated food as it reaches consumers, and 3) losses in nutritional values. To answer these concerns large numbers of animal-feeding studies have been carried out over the past two or three decades in this and other countries of the world. The results seem to have provided convincing evidence that irradiated foods are wholesome and safe for human consumption. Microbial health hazards in these foods are about the same as or lower than for foods preserved using other technologies. Nutrient losses vary with dose level and other factors but can be kept within tolerable limits.

In spite of this evidence, as of December 1981 the FDA had not issued regulations permitting the commercial irradiation of food. However, the FDA has announced that it is considering regulations permitting low-dose irradiation to destroy insects in cereal products, in certain unprocessed fruits, and in dry spices and condiments (FDA, 1981). In addition, this agency announced its intention to provide guidelines for conducting additional animal-feeding studies that may be required to show the safety and wholesomeness of red meats, poultry, and fish irradiation at medium dose levels. No mention was made of issuing regulations for high-dose irradiation of food.

The Codex Alimentarius Commission (an agency of the United Nations concerned with food standards and related matters) has moved somewhat more expeditiously on the international scene. The commission has recommended approval of food irradiation by member nations at low and medium dose levels. Based on these recommendations, Bulgaria, Canada, France, Holland, Hungary, and the Soviet Union have issued regulations permitting the use of radiation for treating a variety of foods, including chicken, fish, certain frozen foods, spices and condiments, and certain unprocessed products such as potatoes, onions, and mushrooms (IAEA, 1980). In passing, it is interesting to note that for a number of years the food regulatory agency of Japan has permitted the irradiation of potatoes to control sprouting.

Just what the future holds for the commercial radiation processing of food in this country is difficult to assess. Even if the FDA finally issues regulations permitting its use, will the U.S. food industry adopt this new technology? To do so it will have to invest large sums of money to construct the special facilities needed for radiation processing. In addition, it may have to finance additional animal-feeding studies to provide safety data FDA may require for medium- and high-dose irradiation. Such studies are expensive and very time-consuming. Whether the economic incentives are there to induce companies to spend the time, effort, and money to begin radiation processing on a commercial basis is a moot question. Only time will tell.

REFERENCES

Ayers, J. C., Mundt, J. O., and Sandine, W. E. (1979). "Microbiology of Foods." Freeman, San Francisco, California.

Bigelow, W. D., and Esty, J. R. (1920). Thermal death point in relation to time of typical thermophilic organisms. J. Infect. Dis. 27, 265–280.

Code of Federal Regulations (1979). "Titles 7, 9, 21, 27, 40, 42 and 50." U.S. Gov. Print. Off., Washington, D.C.

FDA (1981). News release (P81-6), Mar. 27. U.S. Dep. Health Hum. Serv., Washington, D.C.

Heid, J. L., and Joslyn, M. A. (1967). "Fundamentals of Food Processing," Vol. 1. Avi, Westport, Connecticut.

IAEA (1980). News release (PR80-28), Dec. 4. Int. At. Energy Agency, Vienna.

Labuza, T. P., Tannenbaum, S. R., and Karel, M. (1970). Water content and stability of low-moisture and intermediate moisture foods. Food Technol. (Chicago) 24, 543–550.

Loncin, M., Bimbenet, J. J., and Lenges, J. (1968). Influences of A_w on the spoilage of foodstuff. J. Food Technol. 3, 131–142.

Proctor, B. E., van der Graaf, R. J., and Fram, H. (1943). Rep. Quartermaster Contract Proj. July 1, 1942–June 30, 1943, Food Technol. Lab., Mass. Inst. Technol., Cambridge, Massachusetts.

Shepherd, A. D. (1960). Time-temperature–tolerance of frozen foods. U.S. Agric. Res. Serv. ARS 74-21, pp. 18–25.

Stewart, G. F., and Kline, R. W. (1968). Factors influencing rate of deterioration of dried egg albumen. Ind. Eng. Chem. 40, 916–919.

Stuckey, B. N. (1968). Antioxidants as food stabilizers. In "Handbood of Food Additives" (T. E. Furia, ed.), p. 185–223. Chem. Rubber Publ. Co., Cleveland, Ohio.

Stumbo, C. R. (1973). "Thermobacteriology in Food Processing," 2nd ed. Academic Press, New York.

Troller, J. A. (1982). "Sanitation in Food Processing and Food Service." Academic Press, New York.

Troller, J. A., and Christian, J. H. B. (1978). "Water Activity and Food." Academic Press, New York.

Urbain, W. M. (1978). Food irradiation (a review). Adv. Food Res. **24**, 155–227.

Van Arsdel, W. B., Copely, M. J., and Olsen, R. L. (1969). "Quality and Stability of Frozen Foods." Wiley (Interscience), New York.

Wüst, O. (1930). Preserving food. Fr. Patent 701,302.

SELECTED READINGS

Fennema, O. R., Powell, W. O., and Marth, E. H. (1979). "Low-Temperature Preservation of Foods and Living Matter." Dekker, New York.

Furia, T. E., ed. (1973, 1980). "Handbook of Food Additives," 2 vols. Chem. Rubber Publ. Co., Cleveland, Ohio.

Hanson, H. L., Campbell, A., and Lineweaver, H. (1951). Preparation of stable sauces and gravies. Food Technol. (Chicago) **5**, 432–440.

International Commission on Microbiological Specifications for Foods (1980). "Factors Affecting Life and Death of Microorganisms," Vol. 1. Academic Press, New York.

Karel, M., Fennema, O. R., and Lund, D. B. (1975). "Physical Principles of Food Preservation." Dekker, New York.

National Academy of Sciences (1973). "Use of Food Additives in Production, Processing, Storage and Distribution." Natl. Acad. Sci., Washington, D.C.

Reed, C. R. (1975). "Enzymes in Food Processing," 2nd ed. Academic Press, New York.

Chapter 8

Selection of Ingredients and Conversion into Processed Foods

We now turn our attention to the central theme of this text—the selection of raw materials and other ingredients and their conversion into safe, palatable, convenient, stable, and nutritious processed foods. The stage for this has been set by first carefully examining the various elements that should be considered before attempting to manufacture these foods: 1) basic elements of food quality, 2) food safety and principles for control, 3) sensory properties of food and their evaluation, 4) nutritional values of food and their optimization, and 5) quality deterioration and spoilage problems of processed food and principles for control (see Chapters 3–7.) In this chapter we will learn how these are taken into account in the manufacture of foods in general, and we will examine in detail a few typical foods that are produced using the various processing and preservation technologies.

The first part of the chapter will deal with the general principles and practices that are employed in producing processed foods: 1) end-product definition; 2) selection and handling of raw and other ingredients; and 3) the mechanical, chemical, biochemical, and microbiological operations used to transform ingredients into consumer foods (or into intermediate products used in further processing). The second part of the chapter will describe how these principles and practices are currently being used in the manufacture of some common processed foods found in American markets.

Underlying Principles

Product Identity: Ingredients and Formulation

Commercially processed foods vary in complexity from simple items such as canned fruit to the complex array of items found in a frozen TV dinner. In either case we must first identify what specific product is going to be produced, its ingredients, the steps by which it will be prepared, and finally, the means that will be used in its preservation. In order to make it clear to the reader what is involved in this process we will examine two examples. Our first subject is canned fruit, or more specifically, canned peach halves in sugar syrup, packed in an enameled "tin" can (see Fig. 52) and sterilized in a hot-water retort. The product is safe to eat as is, and is expected to have good palatability and nutritional values appropriate for the food group to which it belongs (see Chap. 6). These qualities are expected to be retained essentially unchange for a year or more without refrigeration. It is also very convenient to use, requiring only a can opener, a spoon, and a serving dish. In other words, canned peach halves possess the essential quality attributes for a food of this type. The second example is a frozen TV dinner—specifically, a dinner consisting of roast beef, mashed potatoes and gravy, corn on the cob, and a buttered roll. This is complete main meal lacking only a beverage and dessert. The components are partially processed beforehand (roast beef from chilled beef, mashed potatoes, cooked corn on the cob, gravy from a dry mix, freshly baked rolls, and freshly churned butter) and assembled in a serving receptacle. The receptacle is then crimp-covered with aluminum foil. It is usually packaged using a specially coated paper; after sealing the packaged assembly is quick-frozen. The TV dinner is perfectly safe to eat provided the instructions on the label for heating and serving have been followed. The components are expected to be of acceptable palatability and to possess nutritive values appropriate for the classes of food represented. The quality of the meal should remain essentially unchanged for a reasonable time period (3–9 months), provided the package remains sealed and the food has been kept at a temperature of 0°F (-17.8°C) or lower. This dinner is convenient to use (compared to preparing the meal from scratch), requiring only heating for a few moments in a microwave oven* or somewhat longer in a conventional oven before serving. Thus the TV dinner possesses all of the quality attributes required for good consumer acceptability.

We will discuss other examples of processed foods in more detail later.

---

*If a microwave oven is used, the serving receptacle and cover must be made of a nonmetallic material (usually a coated paper). Aluminum foil absorbs microwave energy extremely rapidly and would soon overheat, perhaps even melt!

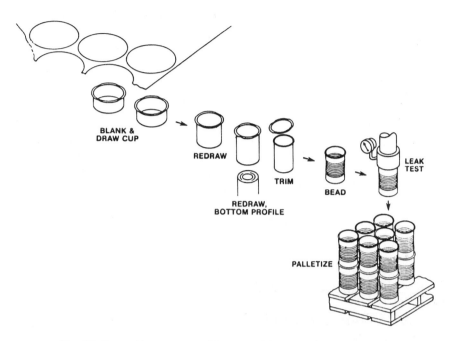

BLANK &
DRAW CUP

REDRAW

TRIM

REDRAW,
BOTTOM PROFILE

BEAD

LEAK
TEST

PALLETIZE

Fig. 52. Can-making processes. (Courtesy of American Can Company.)

Raw-Materials Selection and Handling

Quality control for processed foods requires that proper attention be paid to the raw materials and other ingredients used in their formation. This involves careful inspection and sorting of the fruit, vegetables, livestock, poultry, milk, eggs, and so on, as they are received at the food processing plant. Increasingly it involves control over the genetic, nutritional, and physiological background, the cultural or husbandry practices, and the disease- and pest-control programs used in the production of the raw material and other ingredients. Selection and handling procedures can affect any or all quality attributes. However, palatability and nutritional values are the attributes most commonly affected.

Genetic Factors. Many of the characteristics of the ingredients used in processed food can be favorably affected by controlling their genetic makeup. Recognition of this fact has led to an increase in the attention paid to quality by plant and animal geneticists. Until recently, these specialists concerned themselves mostly with product yields, growth rate, disease resistance, and such things, and gave little or no attention to quality and processing characteristics. Today, however, ever more attention is being paid to sensory quality, nutritional values, and processing characteristics of raw materials for producing processed foods. Breeding special strains of wheat for bread-making is an example of applying genetics to improve quality. Wheat breeders have been able to develop strains that yield flours with greatly improved qualities for bread making. Other strains have been developed for making better pasta products and still others for superior cake-making properties. Plant breeders have produced potato and onion varieties that are superior for dehydration, green pea and broccoli strains and sweet corn hybrids for canning or freezing, peach varieties for canning or freezing, and so forth. Livestock and poultry geneticists have developed strains and crosses that provide products of specific types and qualities. For example, dairy cattle breeds have been introduced that produce milk of a desirable fat content; beef cattle have been bred in which the amount and distribution of meat, bone, and fat in the carcass are controlled. Chicken and turkey crosses have been developed in which body conformation, the size of breast, thigh, and drumstick, and the amount and distribution of flesh are controlled.

Thus it may be seen that genetics can be very useful and effective in the control of the quality of food raw materials. In fact, it is becoming increasingly common for the processors to specify special breeds, varieties, strains, or crosses for particular processing purposes. For example, some producers of precooked frozen chicken insist on a certain broiler cross, certain strains of tomatoes are often specified for processing into juice and paste, specific varieties of potatoes are specific for manufacturing granules and so on.

Physiological Factors. The type and quality of food raw materials are also affected by certain physiological factors. Hormones especially influence the physiological behavior of plants and animals and are capable of producing profound changes in chemical composition and quality. In meat production, castration of the male is an example of physiological treatment that favorably affects quality. Castration changes the hormone balance in the animal, and in cattle this leads to increased fat deposition. The result is meat that is juicier and more tender. Castration of swine and sheep prevents the occurrence of the strong odors associated with boar and ram meat. Synthetic female sex-hormone feeding is sometimes used in broiler-fryer chickens to enhance tenderness and juiciness.

Physiological control for improving the quality of fruits and vegetables is of more recent origin than for livestock and poultry. However, ethylene gas has been used for years to stimulate ripening in certain fruits, such as bananas. Bananas picked in an immature state and shipped to markets are ripened with ethylene before sale. This process makes it possible to ripen bananas as needed, which would not otherwise be possible because of the distance of growing areas (the tropics) from the market and the high perishability of the ripe fruit. Other active chemicals are now available that affect fruit and vegetable quality, such as gibberellic acid, which markedly changes the size and appearance of plant tissues, and is used to control fruit size and appearance. Undoubtedly other physiologically active agents will be discovered for use in improving the quality of fruits and vegetables for processing.

Irrigation and Fertilizing Practices. Although our knowledge of this subject is still limited, it would appear that irrigation and fertilizer may affect the size, appearance, flavor, and nutritive value of fruits and vegetables (sometimes adversely). For example, some scientists claim that excessive soil moisture and high nitrogen fertilizer levels decrease the quality of canning peaches, although this has been disputed. Deficiencies in soil moisture and plant nutrients reduce fruit size. However, much more work needs to be done in this field before we can be sure of good and bad effects. The day will undoubtedly come when the practices for the use of irrigation and fertilizers for fruits and vegetables for processing will be specified (Fig. 53).

Animal Nutrition. The types of feed used as well as feeding methods affect carcass quality of livestock and poultry. For example, high-calorie diets produce increased fatness in broilers; reduced exercise and *ad libitum* feeding produce a fatter beef carcass. Additional work is required along these lines for all types of livestock and poultry. It seems certain that greater control will be exercised by the processor over the practices used by the producer in feeding livestock and poultry for meat production.

Fig. 53. Plant breeders examining broccoli varieties. Those especially suitable for freezing have been developed by seed companies, U.S. Dept. of Agriculture, and state experiment stations. (Courtesy General Foods Corporation.)

Cultural and Husbandry Practices. Such factors as planting schedules, plant spacing, pruning and thinning, and pest-control programs affect fruit and vegetable quality. For example, corn earworms ruin the appearance of corn-on-the-cob for freezing. Air-sacculitis (a disease affecting the respiratory tract) in broilers results in the condemnation of many carcasses and a marked downgrading of others, due to poor fleshing, hemorrhaging, and a darkening of the flesh. Fortunately, a good deal of research has been done on controlling plant and animal diseases and pests. However, there must be a continuous, strong effort made along these lines to maintain an advantage. The food scientist must work closely with plant pathologists, entomologists, veterinarians, and other specialists in assuring good quality raw material for processing.

Climatic Factors. Climate has a marked influence on the composition and quality of many food plants and animals. Thus certain regions of the world are renowned for their wine because of a favorable climate (and soil conditions) for growing the grapes. Climate affects color (e.g., generally more color is developed in a cool region than in a warm one), acidity (higher in a cool region), sugar content (usually higher in a warm region), and so on. Plant and animal physiologists have made great progress in recent years in growing plants and animals in controlled environments, thus making available more information on the climatic factors influencing quality (Fig. 54).

For more information on raw material selection and handling, see Ryall and Lipton (1979), Ryall and Pentzer (1974), and ASHRAE (1978).

Fig. 54. Napa Valley (California) vineyard. Climate and geography are important in production of grapes suitable for wine making. (Courtesy Wine Institute.)

Preparation, Processing, and Preservation Operations

A wide variety of operations are used in transforming raw materials and other ingredients into processed foods. The purpose is to produce processed foods that are safe, palatable, convenient, stable, and nutritious. It is difficult to categorize simply the variety of steps that are used in food manufacture. For example, heat treatment is commonly used not only to destroy microbes and inactivate enzymes but also for cooking. The food scientist must learn the various operations that are used and what they accomplish in the production of quality processed foods. In order to make the task as simple and understandable as possible we have chosen to classify these operations as follows: 1) mechanical, 2) physicochemical and chemical, 3) biochemical, and 4) microbial. We will examine each type and provide some specific examples of their use in food manufacture later on.

Mechanical Operations. Some of the principles and practices used were developed by mechanical and industrial engineers; others have been invented by talented artisans and craftsmen working in the processing plant. Table 30 shows the more important mechanical operations used in food manufacture. Figure 55 is a simplified chart showing product flow through a processing plant, from the raw materials received to the processed food leaving the plant.

Physicochemical Actions and Chemical Reactions. There are many operations in food manufacture that involve physicochemical action and/or chemical reactions. These now usually make use of principles and practices developed by chemists and chemical engineers. Others have been developed by food scientists and by talented artisans. Table 31 shows the more important operations of this type.

As the table shows, a great variety of physicochemical and chemical reactions are used. Two examples of physicochemical reactions are: 1) precipitation reactions, in common use as in the clarification (fining) of wine with gelatin or bentonite to remove cloudiness and sediment; and 2) emulsification of aqueous ingredients and oil, accomplished by the use of an emulsifier and physical treatment, such as egg yolk and homogenization.

Two examples of chemical reactions used in processing and preservation are the bleaching and maturing of flour and the sanitizing of food-contact surfaces of processing equipment. In the first case chemicals, including chlorine dioxide, are used to oxidize the natural pigment (β-carotene) in flour, as well as to hasten the maturation process, which improves bread-making performance. In plant sanitation, chlorine and chlorine-based compounds are widely used to destroy microbes remaining on food-contact surfaces following cleaning operations.

TABLE 30
Mechanical Operations Used in Food Processing

Materials handling	Disintegration	Separation	Quantifying	Combining and Blending	Heat treatment
Conveying (by belt, chain, wire mesh, bucket, screw, etc.)	Crushing	Pressing	Sizing, screening	Mixing	Heating
Fluming (by flowing water)	Pulverizing	Filtering	Weighing, scaling	Blending	Cooking
Pneumatic (by moving air)	Grinding	Sieving	Counting	Homogenizing, colloid milling, ball milling	Blanching
	Shredding	Sifting	Filling (volumetric)	Working	Steam exhausting
	Flaking	Squeezing, reaming		Kneading	Retorting, autoclaving
	Dicing	Sedimenting, clarifying			Pasteurization
	Cubing	Centrifuging			Sterilization
	Milling	Peeling			Steaming
		Stemming			
		Dehairing			
		Defeathering			
		Deboning			
		Skinning			
		Sorting			

Refrigeration	Concentration (partial water removal)	Dehydration	Forming and shaping operations	Surface treatments	Packaging and packing operations
Cooling, chilling	Evaporating, condensing	Sun-drying	Forming	Coating	Filling
Freezing	Freezing/ centrifuging	Tunnel-drying	Shaping	Breading	Sealing
Cold storage		Drum-drying	Rounding	Decorating	Packing, casing
Cooler storage		Spray-drying	Extruding	Enrobing	Strapping, taping
Frozen storage		Bin-drying	Puffing		Palletizing
		Fluidized-bed drying			
		Freeze-drying			

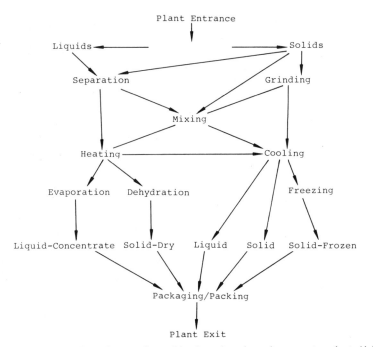

Fig. 55. A schematic chart showing flow of food product through processing plant. (Adapted from Heldman and Singh, 1980.)

TABLE 31
Physicochemical and Chemical Processes of Food

Physicochemical	Chemical
Crystallization	Oxidizing, bleaching, maturing
Flocculation, precipitation, fining	Reducing, hydrogenating
Agglomeration, instantizing	Hydrolysis
Carbonation	Acid hydrolysis
Aerosol formation	Alkaline hydrolysis
Emulsification	Sulfiting
Thickening	Chemical disinfection
Gelling	Chlorination
Coagulating	Iodine compound treatment
Smoking	Quaternary ammonium compound
Cleaning/washing/rinsing	treatment
Churning	Hydrogen peroxide treatment
Laminating, binding, adhering	Chemical leavening
Foaming	Nitrate curing
Whipping, beating	

Biochemical Actions. Many biochemical transformations find application in food processing. They make use of principles and practices developed by biochemists, enzymologists, food scientists, and biochemical engineers; some have been developed by gifted artisans. Table 32 shows some of the more important types.

Biochemical reactions involve the action of naturally present or added enzymes that perform a variety of useful functions in the processing and preservation of food. For example, malt contains large amounts of β-amylase, which, in the malting of barley for beer making, splits the starch of barley and other grains that are present in beer into maltose. Yeast enzymes, which cannot attack starch, can then convert the maltose to glucose and eventually into alcohol and carbon dioxide, two essential ingredients of beer. Lactase is another hydrolytic enzyme that splits the disaccharide lactose (milk sugar) into the monosaccharides glucose and galactose. This reaction is used in the production of lactose-free milk products, which lactose-intolerant persons can assimilate. Proteolytic enzymes are very useful for a variety of purposes. Bromelain and ficin, for example, are used in chill-proofing beer (preventing cloudiness and sediment due to protein precipitation). Rennin, an enzyme that has both protein-splitting and protein-coagulating properties, is in common use in cheese making. Its coagulating action clots milk, and its protein-splitting action is partially responsible for the desired flavor and texture of various cured cheeses.

Microbial Fermentations. Fermentation is defined as a process of anaerobic or partially anaerobic oxidation of carbohydrates. Food fermentation is the conversion of raw materials by controlled biochemical transformation into a more stable food. Fermentation generally produces a product that is less subject to microbial spoilage than the raw material. By removing (or reducing) the amount of unstable constituents present, the fermentable portion is converted into products that are stable and that, in most cases, exist in sufficient concentration to act as a preservative. The most important microbial-inhibiting

TABLE 32
Biochemical Reactions Used in Food Processing

Hydrolysis	Oxidation/reduction	Isomerization	Coagulation
Diastaste action	Glucose oxidase action	Glucose isomerase action	Rennin action
Lipase action			
Lactase action			
Proteolysis			
Pectin-splitting enzyme action			

compounds produced by fermentation are ethanol, acetic acid, propionic acid, and lactic acid. Fermentations are used in making alcoholic beverages, vinegar, cheese, soy sauce, and sauerkraut. Table 33 shows the fermentations used in the food industry.

The preservative effect of fermentation is important, but the success of most fermentation processes obviously depends on the sensory properties of the by-products. Fermented foods are more flavorful than the raw materials used (e.g., cabbage vs. sauerkraut), or the product may have pleasant effects (e.g., beer and wine). The flavor may be so desirable that the fermented food is used to flavor other foods (e.g., vinegar). In some cases the fermentation process at one time may have had nutritional significance. However, modern beers and wines are filtered so they are yeast-free and thereby contain only small amounts of nutrients, except for their energy content.

The microbes that make fermentation possible are a limited number of bacteria, molds, and yeasts. The microorganisms decompose organic constituents to secure energy for their own growth. The actual conversion takes place in a series of steps, each mediated by its own enzymes and other necessary material (certain metals, for example). Small but often significant amounts of by-products may remain after the fermentation. In addition to the principal fermentation, other microbially induced chemical changes may occur with proteins and fats.

Mechanization and Automation

A glimpse at developments that are currently taking place in the food-processing industry of America—the rush into mechanization and automation—will be most revealing. These changes portend a "new age" for the food industry that will lead to lower manufacturing costs and, it is hoped, also to improved quality of the processed foods, at least in uniformity of quality.

We will now take a closer look at some recent innovations. Until recently, many of the methods and even some of the equipment used in food processing differed little from those developed centuries ago. With the tremendous opportunities now being provided by developments in chemistry, metallurgy, polymer science and technology, mechanical and chemical engineering, and electronics, a revolution is taking place—perhaps a major revolution. New and better materials are available for the construction of equipment and facilities; new concepts are emerging for their design and construction; computer-assisted methods are being developed for designing, fabrication, and so on—all leading to food-processing and preservation systems that barely resemble those of a decade or two ago. One is impressed by these advances on visiting the new food-processing plants around the country (e.g., the modern edible-oil refinery, flour mill, milk processing plant, food dehydration plant, etc.).

TABLE 33

Microbial Fermentations Used in the Food Industries[a]

I. Lactic acid bacteria
   A. Vegetables and fruits
      1. Cucumber → dill pickles, sour pickles, salt stock
      2. Olives → green olives
      3. Cabbage → sauerkraut
      4. Turnips → sauerrüben
      5. Lettuce → lettuce kraut
      6. Mixed vegetables, turnips, radish cabbage → paw tsay
      7. Mixed vegetables in Chinese cabbage → kimchi
      8. Vegetables and milk → tarhana
      9. Vegetables and rice → sajur asin
      10. Dough and milk → kishk
      11. Coffee cherries → coffee beans
      12. Vanilla beans → vanilla
      13. Taro → poi
   B. Meats → sausages such as salami, Thuringer, summer pork roll, Lebanon bologna, cervelat
   C. Dairy products
      1. Sour cream
      2. Sour milk drinks—yogurt, cultured buttermilk, skyr, gioddu, leban, dadhi, taette, mazun
      3. Butter—sour-cream butter, ghee
      4. Cheese—Unaged → cottage, pot, Schmierkäse, cream whey → mysost, primost, ricotta, Schottengisied
         Aged → cheddar, American, Edam, Gouda, Cheshire, provolone

II. Lactic acid bacteria with other microorganisms
   A. Dairy products
      1. With other bacteria
         a. Propionic acid bacteria → Emmenthaler (Swiss), samsoe, Gruyère cheeses
         b. Surface-ripening bacteria → Limburger, brick, Trappist, Müenster, Port du Salut
      2. With yeast → kefir, kumiss
      3. With molds → Roquefort, Camembert, Brie, hand, Gorgonzola, Stilton blue, Oregon blue, etc.
   B. Vegetable products
      1. With yeasts → nukamiso pickles
      2. With mold → tempeh, soy sauce

III. Acetic acid bacteria—wine, cider, malt, honey, or any alcoholic and sugary or starchy products may be converted to vinegar

---

TABLE 33 (continued)

IV. Yeasts
   A. Malt → beer, ale, porter, stout, bock, pilsner
   B. Fruit → wine, vermouth
   C. Wines → brandy
   D. Molasses → rum (with some flavor contribution from bacteria)
   E. Grain mash → whiskey
   F. Rice → sake, sonti
   G. Agave → pulque
   H. Bread doughs → bread

 V. Yeasts with lactic acid bacteria
   A. Cereal products → sourdough bread, sourdough pancakes, rye bread
   B. Ginger plant → ginger beer
   C. Beans → vermicelli

VI. Yeasts with acetic acid bacteria
   A. Cacao beans
   B. Citron

VII. Mold and other organisms
   A. Soybeans → miso, chiang, su fu, tamari sauce, soy sauce
   B. Fish and rice → lao, chao

ªAdapted from Heid and Joslyn (1967).

The modern digital computer is making automation a reachable goal for many plant operations. It is now possible, using the computer along with appropriate sensing and control instrumentation, to monitor operating conditions; if they are not optimal, the system can be made to "command" necessary adjustments. The requirements for unskilled labor to carry out plant operations are thus greatly minimized by automation. However, the requirement for well-trained and experienced professionals to supervise them is increased, although their numbers need never be very large. This revolution seems to be in its infancy and we would expect to see it continue as long as productivity can be increased, costs lowered, and the uniformity and quality of the product improved.

Selected Commodity Technologies

Using some typical technologies for processed foods as examples, we will next examine how the foregoing principles are currently being applied to food manufacture in the United States. Particular emphasis will be placed on the key preservation techniques in common use: 1) heat treatment, 2) refrigeration, 3) water-activity control, 4) fermentation, 5) chemical preservation, and 6) combinations of two or more of these.

Refrigerated Foods—U.S. Choice Beef

The fresh beef found in the market is an excellent example of a food that owes its shelf life primarily to refrigeration (usually at temperatures a few degrees above the freezing point) and to proper packaging. Whereas the shelf life of beef is limited (a few weeks for wholesale cuts and a few days for retail cuts), tens of millions of pounds are marketed every week with very little deterioration or spoilage.

Raw Materials Selection and Handling. Selected breeds and crosses of cattle are used for the production of high-quality beef. Hereford and Black Angus cattle are popular for this purpose, but increasingly other breeds and cross-bred animals are being used. Beef herds are reared to several hundred pounds live weight using a variety of systems, including the traditional practice of open-range grazing. The cattle are then usually transferred to semiconfinement quarters for "finishing" (fattening). This period varies from 3 to 9 months, depending on the time required to produce an animal that will grade out U.S. Choice after slaughter, evisceration, and chilling. Animals ready for slaughter are commonly shipped by truck to the meat packing plant, using handling and shipping practices that minimize stress caused by high temperatures, excitement, and excessive activity. A stressed animal yields a poor-quality carcass due to off-color and inferior texture.

Slaughter and Processing Operations. Packing plant operations commence with the slaughter of the animal using an approved humane method. After bleeding, the head and hooves are removed and the animal is skinned. The carcass is then opened and viscera withdrawn to permit inspection for wholesomeness by U.S. Department of Agriculture Veterinarian or his designated lay inspector, and for further processing. Carcasses passing inspection are then completely eviscerated and the organ meats separated and placed in a chill room. The carcass is then split in half, trimmed of any inedible tissue, and placed in a cooler for rapid chilling. The chilled carcass is kept under refrigeration [~ 40°F (4.4°C) and at high humidity] for "aging" (tenderizing) for about 6 to 10 days. During this period U.S. Department of Agriculture personnel grade the carcass. Grade is based on external fat cover, degree of intramuscular fat, fat color, color, texture of the musculature, and other factors. U.S. Choice is the second highest quality grade given. Recently there has been a trend toward cutting the carcass halves at this stage into wholesale units (so-called primal and subprimal cuts), and vacuum-packing them in special moisture-vapor-proof, oxygen-impermeable plastic bags for aging and distribution to the supermarket. At the retail market the carcass halves, quarters, or specially packaged wholesale cuts are placed in a cooler until ready for the preparation of the retail cuts. After the latter (steaks, roasts, etc.) have

Fig. 56. U.S. Department of Agriculture grade marking of beef. (Courtesy U.S. Department of Agriculture.)

been prepared, they are packaged in an oxygen-impermeable, moisture-vapor-proof plastic film and placed in a refrigerated display counter.

Current Quality Control Problems. The processing and marketing of U.S. Choice beef presents few quality problems, provided proved industry practices have been followed. However, refrigeration temperatures, holding times, and packaging practices are all important to success. If the meat is held in the display counter too long, the bright red color changes to an unattractive gray or grayish brown. If held an excessively long time the meat will become slimy and smell "sour" due to microbial action.

Heat-Sterilized Foods—Canned Green Peas

Canned green peas are a staple vegetable in the American diet, representing a good example of a processed food that owes its excellent shelf life to proper packaging and heat sterilization. It is a member of the family of so-called shelf-stable foods—a term food scientists use for foods that keep well at ambient (room) temperatures for a year or more without showing significant deterioration or any evidence of spoilage. Such foods are expected to be free of viable microbes that cause infections or intoxications in humans. Canned peas are processed in the following manner.

Raw Materials Selection and Handling. The following practices must be carefully controlled in order to produce high-quality canned peas with a good green color and a succulent, sweet, and tender flavor. Special canning varieties of peas have been developed by plant breeders and are in widespread use. In addition, a cool, moist climate and certain type of soil are required, conditions that are met in parts of such states as Wisconsin, Minnesota, and Oregon. Canning peas are harvested at an early stage of maturity. The pods, then the peas, are separated by devining and dehulling machines in or near the fields in which the peas are grown. They are then quickly trucked to the cannery and chilled with cold water.

Processing and Preservation Steps. The first step taken at the factory is to clean the shelled peas thoroughly, using such means as metal screens and aspirators. They are then mechanically sorted for size. Hot-water blanching is the next operation: this inactivates the enzymes present, removes trapped air (i.e., oxygen), and slightly softens the peas, which makes for more compact can filling later. Blanching conditions consist of heating for about 3 minutes in water at 190° to 200°F (87.8° to 93.3°C). The operations are followed by quality grading using gravity separation in a 9.5–10.0% salt (sodium chloride) brine. The more tender, succulent peas float in the brine and are easily collected from the overflow stream coming from the grader. The adhering brine is washed off the graded peas with fresh water. After visual inspection and sorting (to remove belmished peas), the peas are mechanically filled into cans. Tin-plated cans coated on the inside with a special enamel are usually used for this purpose; the enamel serves to prevent discoloration of the peas due to iron sulfide formation during heat processing and subsequent storage. The peas are then covered with a salt-sugar solution, the ratio of salt to sugar being adjusted to provide an optimal flavor balance. The filled can is then mechanically sealed with a lid (usually of the same composition as the can) under conditions of flowing steam or in a steam/vacuum-closing system. The conditions for sealing are so controlled as to provide a vacuum (7–10 inches of mercury) in the head-space of the canned peas after heat sterilization and cooling. Sterilization is usually carried out using steam under pressure in a stationary or mechanically agitated steam or hot-water retort. The proper operating conditions must be carefully calculated and tested previously, using data collected for thermal death times and heating and cooling rates (see Chapters 4 and 7). Table 34 shows the process times recommended for brine-packed peas, or peas and carrots, in a stationary steam retort, for various can

TABLE 34

Recommended Processing Times for Brine-Packed Peas or Peas and Carrots[a]

Can size	Maximum drained fill weight (oz.)[b]	Minimum initial temperature °F[c]	Minutes at retort temperature 240°F (115°C)	245°F (117°C)	250°F (121°C)
211 × 304	5.9	70	31	20	13
		140	28	18	12
211 × 304	6.4	70	34	22	16
		140	30	20	13
303 × 406	11.5	70	44	32	24
		140	39	26	19
303 × 406	13.5	70	65	51	43
		140	56	43	35
603 × 700	72.0	70	57	40	28
		140	48	32	21

[a]From National Food Processors Association (1977).
[b]Exceeding the fill weight critical factor for a scheduled process constitutes a process deviation and will require an evaluation by a competent process authority to determine if a potential health hazard exists. The processes listed with some of the higher fill weights for a can size may result in a lower-quality product.
[c]70°F = 21.1°C; 140°F = 60°C.

sizes, fill weights, and sterilization temperatures. The cans are rapidly cooled using water sprays. Paper labels may be attached any time after the cans are thoroughly dry; however, if lithographed cans have been used, this step is obviously unnecessary.

Current Quality Control Problems. The processing and preservation technology used for canned green peas has been so refined over the past two or three decades that few quality control problems are encountered, especially if the generally accepted manufacturing practices have been followed.

Frozen Foods—Orange Juice Concentrate

Frozen concentrate of orange juice is an example of a food that owes its excellent shelf life to the use of low temperature for freezing and storage and distribution [0°F (−17.8°C) or lower]. Properly packaged, it will retain its sensory quality and nutritive values for a year or more. Because of these facts and because orange juice concentrate is reasonable in price and popular, many hundreds of thousands of tons of oranges are processed each year into frozen orange juice concentrate in this country.

Raw Materials Selection and Handling. Special varieties and certain climatic and soil conditions are required for the production of oranges suitable for processing into frozen concentrate. Florida, Texas, and California produce almost all of the oranges used for this purpose, with the former dominating the market. Florida oranges possess the highest sugar-to-acid ratio in the juice, providing optimal sweetness without the necessity of adding sugar. Much of the fruit is picked by hand, but satisfactory mechanical harvesting methods are gradually being developed. (Damage to the trees and injury to the quality and processing characteristics of the fruit are the major problems.) The harvested fruit is collected in bulk bins and trucked to a nearby processing plant. The fruit may be stored under refrigeration for a short period of time before processing.

Processing and Preservation Steps. The oranges are first examined by a qualified inspector for juice quality and suitability for processing. (A sample is taken from each truckload, extracted, and the juice tested for conformance to state standards, primarily sugar-to-acid ratio.) Oranges from different sources are often blended to provide for better uniformity of quality in the final product (e.g., high-acid fruit is blended with low-acid fruit).

The first step in processing is to wash the fruit thoroughly using rotating brushes and water sprays. The oranges are then juiced automatically using an extractor such as illustrated in Fig. 57. The juice is next concentrated in a multistage vacuum evaporator. Enzyme inactivation by heat [at about 210°F (98.9°C)] is accomplished in the first stage of evaporation. This step is required to avoid gelation (thickening or gel formation) in the juice later during processing and packaging. Concentrated juice comes from the evaporator at about 50–65% solid but is diluted back with fresh juice to about 45% solids.

Juice essence (aroma fraction) is recovered from the vapors leaving the evaporator and is concentrated in a separate fractionating column. The essence plus about 0.25% peel oil and some fresh juice are next blended into the concentrate. These "add-backs" are used to give the final product, when reconstituted, a flavor balance similar to that of freshly squeezed orange juice. The product is next cooled to about 30°F (−1.1°C) and packed into barrels with plastic film liners. The concentrate from a number of barrels may be blended to provide a more uniformly flavored final product. Most of the concentrate is then packaged in 6-oz. consumer-size cans, which are made up of a plastic-coated, paperboard body with coated metal ends. The packaged product is usually frozen in an air-blast tunnel at −40°F (−40°C) or below. The product is stored and distributed at 0°F (−17.8°C) or lower.

Current Quality Control Problems. Generally speaking, there seem to be few serious quality control problems in the manufacture of frozen orange juice concentrate. However, periodic unfavorable weather, especially heavy frosts in the orange groves, makes it difficult to maintain a uniform flavor balance and, of course, can create a shortage of fruit for processing.

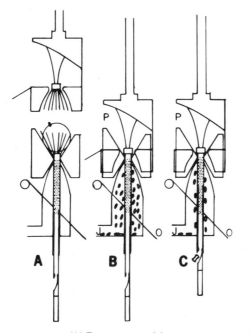

Fig. 57. Extractor for orange juice. (A) Fruit positioned for juice extraction. (B) Primary juicing step. (C) Final squeezing step, using the rising perforated tube. Shredded peel is ejected at P and peel oil fraction is removed at the inclined plane marked O. At J, juice flows out of extractor. (Courtesy FMC Corporation.)

Heat-Treated, Chilled Foods—Grade A Pasteurized Milk

Grade A pasteurized milk is an example of a food that owes it microbial safety and shelf life to a combination of heat treatment and refrigeration (slightly above freezing). Pasteurized milk has a limited shelf life [10] days plus at 35°–40°F (1.7°–4.4°C)]. However, current methods for marketing and distributing the product in the United States make it possible to market billions of gallons of milk per year with little risk of deterioration or spoilage. The steps used in processing beverage milk are more complex than most people realize. They are summarized as follows.

Raw Materials Selection and Handling. Beverage milk is produced chiefly by herds on dairy farms located fairly close to population centers. Several breeds of dairy cattle are used, but in this country the Holstein breed predominates because it is a very efficient milk producer and possibly also because the fat content of the milk is close to the legally required amount. Dairy animals in milk production are commonly kept in confinement and are fed scientifically balanced rations. Besides its nutritional benefits, controlled feeding prevents the production of off-flavored milk due to the animals consumption of certain noxious weeds.

Great stress is placed on sanitation in the housing and handling of dairy cattle, especially during milking operations. First and foremost, the dairy animals must be free of diseases communicable to humans. The workers caring for the animals must also be healthy and possess sanitary work habits. Finally, the milking environment and the equipment used must be kept scrupulously clean and sanitary.

Contrary to practices used for most other foods, the grading of fluid milk is determined primarily by conditions found at the point of production. In the United States, milk is graded on the basis of its acceptability for consumption 1) as a fluid product (Grade A) or 2) for the manufacture of dairy products (Grade B). Each state has the responsibility for promulgating and enforcing specific regulations for these grades of milk sold within its borders. However, for all dairy products entering into interstate commerce, federal laws and regulations apply (see Chapter 9). The raw milk is rapidly chilled after milking and is kept under refrigeration [45°F (7.2°C) or below] until ready for delivery to the processing plant.

Processing and Preservation Steps. Chilled raw milk is customarily delivered in sanitary insulated tank trucks to the plant, where it is transferred to a refrigerated holding tank pending processing. The milk is first filtered and centrifugally clarified to remove any objectionable particulate matter. It is then standardized for fat content (a 3.5% fat content is the minimum usually specified, but it may vary from 3.6 to 4.0%). Milk fat (as cream) is separated from the nonfat portion using a centrifugal cream separator. (The cream and skim milk are used for the standardization.) The milk is next homogenized to reduce fat globule size to a point where a cream layer will not form on holding. It is then pasteurized, using one of three approved processes (established from thermal death studies carried out on the most heat-resistant pathogenic microbes found in raw milk): 145°F (62.8°C) for 30 minutes; or 161°F (71.7°C) for 15 seconds; or 280°F (137.8°C) for at least 2 seconds. The third heat treatment (commonly termed high-temperature/short-time pasteurization, or HTST) has gained favor commercially; not only does it destroy all pathogens, but it also provides a better shelf life with minimal changes in flavor and color. Pasteurized milk is rapidly chilled to 45°F (7.2°C) or lower, filled into quart or half-gallon plastic-coated paperboard cartons, and heat-sealed. However, gallon-size polyethylene jugs sealed with a metal closure are being used more and more often. Grade A pasteurized milk is now ready for delivery at less frequent intervals (e.g., once or twice a week).

Current Quality Control Problems. The dairy industry has done a magnificent job of carrying out the research required for producing Grade A pasteurized milk. This is especially true in assuring the consumer of a product that is safe and pure and possesses good sensory properties, a reasonable shelf life, and excellent nutritional values. Few quality control problems exist in the industry providing proved practices have been followed. However, from time to time certain off-flavor problems plague the industry (e.g., those due to lipid oxidation).

Dry Food Products—Vacuum-Packed Coffee

A good example of a food (beverage) that owes its shelf life to a very low water activity (A_w) and to a special packaging method is vacuum-packed coffee. The A_w is extremely low, well below 0.60, the minimum required by spoilage microbes. The vacuum packing reduces the oxygen content in the can of coffee to a point that substantially reduces the loss of aroma and any off-flavor development, even at ambient temperatures. As a result, millions of pounds of vacuum-packed coffee are produced each week and sold to consumers with no possibility of any spoilage and with the assurance that, after opening the can, a satisfactory aroma and flavor will be retained for a week or so at room temperature.

Raw Materials Selection and Handling. Green coffee beans are produced and partially processed in a number of tropical countries. However, most of the coffee consumed in the United States comes from Latin America: Brazil, Colombia, Venezuela, Central America, and Mexico. The coffee trees are grown in mountainous areas, usually above 2000 feet elevation. The fruit is hand-picked when it reaches a deep red, cherry-like appearance. In Brazil the harvested coffee berry (as the fruit is called) is dried in the sun or by a combination of sun-drying and mechanical dehydration. In the other Latin American countries, the harvested berries are first allowed to undergo a natural fermentation to help remove the skin and the underlying pulp. They are then washed and sun-dried. The moisture content of the dried green beans, in either case, is about 11%. Next, the beans are stripped of their outer layers, including the endocarp (called parchment), using mechanical devices. Before export the beans are sorted and graded for size, density, and sensory quality. The so-called soft-sweet and clean beans are considered to be of the highest quality and usually bring a premium price. Trade names for the green beans indicate the country of origin and production locale (e.g., the Manizales coffee of Colombia and the Santos, Parana, and Minas coffees of Brazil).

Further Processing and Preservation Steps. Based on the sensory evaluation of the brewed coffee made from a number of individual lots of green beans, a special blend is made up for each name brand of ground roasted coffee.

These operations are carried out carefully and systematically, not only to obtain a certain type and level of quality, but also to assure uniformity. The blended green beans are then roasted, on a batch basis or continuously, in air heated to 600°–800°F (316°–427°C) for about 15 minutes. The roasted beans are then quickly cooled in air or by a combination of water sprays and air. During roasting, coffee develops its characteristic color, aroma, and flavor. Most of the moisture is driven off; the sugars present are caramelized; and during the final stage, pyrolysis (thermal degradation) of the carbohydrates, proteins, and other constituents takes place. (Gases (mostly carbon dioxide), certain vapors, and sublimates (e.g., part of the caffeine) are driven off during pyrolysis. After roasting and cooling, the beans are ground. The degree of fineness varies with the method used for brewing the coffee (e.g., regular, fine, drip, electroperk, etc.) The last manufacturing step is to fill the coffee into tin-plated steel cans (holding 1, 2, or 3 pounds). The covers are put on and then vacuum packed. Sealing conditions are so adjusted that no more than 1% oxygen remains in the gas surrounding the coffee in the can after sealing.

Current Quality Control Problems. Overall, the industry in the United States seems to encounter few quality problems in the manufacture and distribution of vacuum-packed ground roasted coffee. If proper care has been taken in the selection and blending of the raw beans, in roasting, cooling, and packaging, and by providing for a fairly rapid turnover of stocks in the market, few problems seem to arise. However, if turnover is not practiced or is haphazard, or if stock is left on the retail shelves for more than a few weeks, off-odors and off-flavors may develop due to oxidation. This problem is more serious once the can is opened; the characteristic aroma and flavor of freshly roasted coffee soon disappears and off-flavors become apparent.

Dehydrated Foods—Dried Egg Whites

Dried egg white (commercially known as egg white solids or egg albumen) is an example of a food that owes its shelf life primarily to a low water activity. However, to prevent deterioration one of its components (glucose) must be removed during processing to prevent nonenzymatic browning (see p. 180). Microbial safety of the product is assured by strict sanitation and special heat treatments. Properly prepared and packaged egg white solids have an exceptional shelf life; the product keeps indefinitely at ambient temperatures. Hundreds of thousands of pounds of egg white solids are manufactured and sold each year in the United States for use as an ingredient in candies, cake mixes, meringues, cake toppings, and dietary foods.

Raw Materials Selection and Initial Processing. Shell eggs for the production of high-quality dehydrated products should be not more than a few days old and should possess shells that are sound and have little or no surface contamination. Liquid egg is produced from shell stock in a series of specialized operations. The shell eggs are first placed on conveyors that carry them through grading, washing, and sanitizing steps. Here the soiled and cracked eggs are removed; the remainder are washed using brushes and warm water sprays containing a detergent. After washing, the eggs are sanitized with chlorinated water and then drained. The sanitized eggs are next conveyed to an automated breaker/separator, which opens the egg, removes the shell, and separates the contents into whites and yolks (see Figure 58). During this operation the contents are inspected by an operator. Yolks or whites showing abnormal appearance or an off-odor are removed and discarded. The separated yolks and whites are each then mechanically blended, filtered through a fine screen, rapidly chilled to below 45°F (7.2°C), and finally placed in refrigerated holding tanks for further processing.

Further Processing and Preservation Steps. The first step in processing whites is to pasteurize them at 142°F (61.1°C) for 3–4 minutes. After cooling they are treated to remove the small amount of free glucose (0.4%) naturally present. The glucose is removed by means of either microbial fermentation

Fig. 58. Egg-breaking and -separating machine. (1) Eggs ready for loading; (2) vacuum-loading of eggs; (3) whites separated from yolk; (4) removal of yolk.

or an enzyme, glucose oxidase. No more than 0.05% glucose should remain in the whites after treatment. This process can be accomplished within a few hours' time at 100°F (37.8°C). The liquid is then rapidly chilled to below 45°F (7.2°C). The liquid is next spray-dried in a specially designed chamber using high-pressure atomization into a hot-air stream [300°–350°F (149°–176°C)]. Most of the dried product collects on the floor of the drying chamber, where it is rapidly removed by special conveyors. The remainder is recovered from the exhaust air by means of bag filters made of finely woven textile material (see Fig. 59). Drying conditions for egg white solids must be carefully controlled to avoid heat and mechanical damage to the product and to obtain the requisite moisture content (about 8.0%). The dried product is then put through a vibrating screen to remove any oversize particles. The sieved powder is usually packed and sealed in polyethylene-lined fiberboard drums or boxes. The final processing step consists of heat-treating the packaged product for a week or more at 140°F (60°C). This treatment drastically reduces the total microbial count and destroys most if not all of the salmonella organisms (see p. 92) remaining in the product after spray draying.

Current Quality Control Problems. The production of a salmonella-free egg white solid (which is required by federal regulation) requires a great deal of vigilance and constant attention to detail by operating and quality control personnel. Good raw material, excellent sanitary practices, and strict attention to the conditions of spray drying and heat treatment are required. There

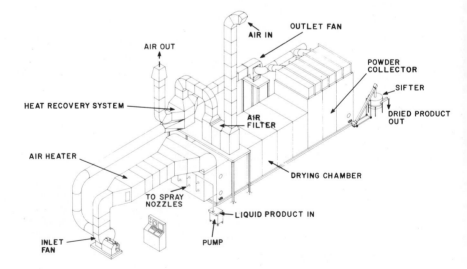

Fig. 59. Spray-dryer used for egg white solids. (Courtesy Henningsen Foods, Inc.)

seems little doubt that the production of salmonella-free egg white solids is one of the most demanding quality control jobs for the food scientist in the food industry.

Foods Preserved by Chemicals—Carbonated Lemon Drink

A carbonated soft drink, such as a lemon-flavored beverage, is a good example of a food (beverage) that owes its shelf life primarily to chemical preservatives and good sanitation in the processing plant. These products have a reasonably good shelf stability (sensory properties and lack of spoilage) at ambient temperatures, perhaps from a few weeks up to a few months. However, distributors usually deliver soft drinks to the markets on a weekly basis, perhaps for economic reasons.

Ingredients Selection. Carbonated soft drinks come under federal regulations as standards of identity products. This means that only certain ingredients are permitted. In the case of a lemon-flavored beverage, these ingredients are potable water, carbon dioxide, sweetener (sucrose, glucose, fructose, sorbitol, or a combination thereof), acidifying agent (citric acid), flavoring agent(s) (lemon juice and lemon peel), and coloring agent (none or tartrazine). Potable water is usually obtained from a municipal water supply but is usually further purified by using a water softener and a special filter to remove calcium and magnesium salts, extraneous solids, microbial agents, and any objectionable flavors. Carbon dioxide is available commercially in liquid form (in steel cylinders under high pressure) or in insulated trucks.

Processing and Preservation Steps. The first step in the manufacture of a lemon-flavored beverage is to prepare a "simple syrup" made up of the sweetener(s) and most of the water required. Citric acid may be added at this point or during the next step. After filtering the syrup, tartrazine color (if used), sodium benzoate, lemon juice, lemon peel oil, and citric acid (if not already present) are added. The carbonation process consists of dissolving carbon dioxide into the formulation under pressure and temperature conditions that result in the finished beverage's containing about 1.5 volumes of the gas per volume of beverage. While still under pressure, the lemon drink is filled into glass or plastic bottles or metal cans and hermetically sealed. Nonreturnable glass bottles and internally coated tin-plated steel cans were in common use until recently. Now, however, returnable glass bottles, plastic bottles, and recyclable aluminum-alloy cans are more common.

Current Quality Control Problems. The carbonated beverage industry currently seems to have few serious quality control problems. This has come about because of the extensive research carried out cooperatively by members of the industry over the years, which resulted in identifying quality problems and working out satisfactory solutions. Of foremost importance are ingredient water quality and sanitary practices in the factory, including the proper washing and sanitizing of containers. In addition, fairly rapid turnover of stocks in the market plays a part in preventing flavor deterioration in certain soft drinks.

Fermented Foods—White Table Wines

Wine is produced by the fermentation of crushed grapes. The variety of grape, the climatic conditions under which it is grown, the time of harvesting, and variations in processing account for wide differences in the type of wine and its flavor and quality. The best quality table wines (less than 14% alcohol) are produced in regions where the grapes ripen sufficiently but where the fruit retains adequate acidity. Regions of low humidity during ripening are best. This limits grape growing and wine making to a narrow band in the northern and southern temperate zones. Within these regions wide variations in the potential quality of the wines that can be produced depend on seasonal variations and such weather conditions as humidity, hail, spring and fall frosts. It is generally agreed that the best table wines are produced in the cooler parts of the temperate zone—areas where grapes ripen with sufficient sugar and not too much acid to produce a balanced wine (20% sugar and less than 1% acid). Areas with low rainfall or humidity during the ripening period produce the best-quality grapes (free of mold) and hence the best-quality wines.

Raw Materials Selection. White grapes are normally used for the production of white table wines. However, red grapes with noncolored juice can also be used. More than 2000 varieties of white wine grapes are known. The wine maker not only requires a variety that has the requisite composition but also one that is free of fungal infection. Moreover, the variety must have a desirable flavor. Among the most desirable varieties are as follows (sold under the varietal name unless otherwise noted): Chenin blanc (Loire), Gewürztraminer, Sauvignon blanc (Fumé blanc, Graves), Sémillon (Sauternes), White Riesling (Moselle, Rhine). Of course, varieties of lesser quality are often used. In the United States these are often sold as California (New York, etc.) chablis or rhine.

Grapes must be harvested when they are at optimum maturity for the type of wine for which they are to be used. For a given variety and moderate temperature this may range over several weeks, but for a specific type it may not last more than a week, especially under warm climatic conditions such as those in California. This is one of the most important command decisions of the grape grower and winemaker.

Grapes are picked by hand but in recent years mechanical harvesters have become available. These have the advantage of more rapid harvest but because of bruising of the fruit it must be transported to the winery in only a few hours. Most grapes are transported in gondola trucks.

Further Processing and Microbial Fermentation. On arrival at the crushing shed of the winery the grapes are crushed and stemmed in rotary crushers operating at high speeds. The juice is separated from the skins by draining (often mechanically) and by pressing. Several types of presses are used but screw-type presses are now common.

If the juice is cloudy (and this often occurs when the fruit is infected by fungi) it may be settled for 24 to 38 hours at as low a temperature as practicable [40° to 50°F (4.4° to 10°C)]. At least 100 mg/liter of sulfur dioxide are added to reduce yeast activity. Alternately, the cloudy juice may be clarified by centrifugation.

To the original or clarified juice, 100 mg/liter or more of sulfur dioxide is added (if no previously used). An *active* pure culture or a desirable strain of *Saccharomyces cerevisiae* is then usually added, especially at the start of the season. Nowadays, commercial pressed yeast of desirable strains is used.

Fermentation ensues with 24 hours. For quality white table wines the temperature is controlled to not over 50°F (10°C). The fermentation may last 3 to 6 weeks. The wine maker normally prefers that the fermentation go to completion (e.g., to less than 0.2% reducing sugar). However, a sweet table wine may be the objective, in which case the fermentation must be stopped before all the sugar has fermented. This is accomplished by reducing the temperature, adding sulfur dioxide, filtering, and centrifuging. French Sauternes, many German wines, and late-harvested California wines are examples of sweet white table wines. (See Amerine and Singleton, 1977, for more detail.)

Stabilization and Finishing. New wine is a cloudy, yeasty supersaturated solution of potassium acid tartrate. Much of the yeast will settle within a few weeks after fermentation ceases if the temperature is not too high [below 50°F (10°C)]. Clarification achieved by racking the wine off the yeast sediment, by chilling, by addition of a clarifying agent such as bentonite, by centrifugation, and by filtration. Tartrate stabilization is accomplished by chilling the wine [to 25°F (−4°C)] and holding for 2 weeks or more. The wine is then cold-filtered off the precipitated tartrate.

White table wines are given nominal aging, usually in large metal tanks, if they will not profit by aging in wood. Modern practice is to age in the wood only wines of higher flavor. and alcohol content, and then not more than 6 to 12 months. Enologists differ in their preference for aging white wines in the wood but the practice is now less in favor.

Prior to bottling the wine will receive a final blending and stabilization. Blending may be done for equalization of the quality of the same wine in several containers or for improving the quality by blends of different wines. Trial bottlings of the final blends are made for stability under hot and cold conditions. Additional and special clarifying agents may be used. The sulfur dioxide will be adjusted to the desired level. If the wine contains fermentable sugar, sorbic acid as well as sulfur dioxide may be used.

The wine is then bottled. In some cases hot-bottling may be employed for wines containing sugar, to ensure stability. However, most such wines are germ-proof-filtered through membrane filters. Close technological control is required.

Following bottling, white table wines lose some of their yeasty and sulfur dioxide odor and thus improve in quality. Some consumers prefer white table wines within a year of bottling. Some of the sweet and more alcoholic types may improve in quality for several years. Eventually the wines darken in color and become oxidized in flavor.

Fermented Foods—Other Wines

Red table wines are produced from red grapes by procedures similar to those already described, except that they are fermented on the skins for 3 to 6 days and are more likely to be aged in wood. Sparkling wines are produced by refermenting a white, pink, or red wine in a closed container with the requisite amount of added sugar to produce a calculated pressure of carbon dioxide. The process is lengthy and labor-intensive, especially in the small containers. The container may vary from a 750-ml container to one of 1000 hectoliters (26,000 gallons), one reason for the higher cost of sparkling wines. In a few cases carbon dioxide may be added directly, but the sparkling wine so produced must be labeled "carbonated." Carbonated and sparkling wines are highly taxed in this country—another reason for their higher cost.

Current Quality Control Problems. The variable quality of the grapes resulting from unfavorable weather conditions remains a major quality problem, especially in cooler grape-growing regions such as Austria, France, and Germany. The technology of crushing, fermentation, and stabilization can prevent major quality control problems but the fact remains that the wines of some years are better than those of other years in spite of the best enological care. Wine making thus retains some artisan characteristics. Enologists do exercise control over the major quality variables by their care in selecting and harvesting grapes, in crushing, fermenting, stabilization, blending, and aging. Because of this the average quality of the wines on the market today is higher than in the past, especially in the United States.

REFERENCES

Amerine, M. A., and Singleton, V. L. (1977). "Wine: An Introduction," 2nd ed. Univ. of California Press, Berkeley and Los Angeles.

ASHRAE (1978). "ASHRAE Guide and Data Book, Application Volume." Am. Soc. Heat., Refrig. Air-Cond. Eng., New York.

Heid, J. L., and Joslyn, M. A. (1967). "Fundamentals of Food Processing." Avi, Westport, Connecticut.

Heldman, D. R., and Singh, R. P. (1981). "Food Process Engineer," 2nd ed. Avi, Westport, Connecticut.

National Food Processors Association (1977). Peas or peas and carrots, in brine. Letter to canners, March 25, 1977. Nat. Food Processors Assoc., Washington, D.C.

Ryall, A. L., and Lipton, W. J. (1979). "Handling, Transportation and Storage of Fruits and Vegetables. Vol. 1 (2nd ed.): Vegetables and Melons." Avi, Westport, Connecticut.

Ryall, A. L., and Pentzer, W. T. (1974). "Handling Transportation and Storage of Fruits and Vegetables. Vol. 2: Fruits and Tree Nuts." Avi, Westport, Connecticut.

SELECTED READINGS

Desrosier, N. W., and Tressler, D. K. (1972). "Fundamentals of Food Freezing." Avi, Westport, Connecticut.

Dunkley, W. L. (1980). Milk and Milk Products. In "Animal Agriculture" (H. H. Cole and W. N. Garrett, eds.), p. 82–103. Freeman, San Francisco, California.

Jackson, J. M., and Shinn, B. M. (1979). "Fundamentals of Food Canning Technology." Avi, Westport, Connecticut.

Johnson, A. H., and Peterson, M. S. (1974). "Encyclopedia of Food Technology." Avi, Westport, Connecticut.

Lawrie, R. A. (1979). "Meat Science," 3rd ed. Pergamon, Oxford.

Loncin, M., and Merson, R. L. (1979). "Food Engineering: Principles and Selected Applications." Academic Press, New York.

Luh, B. S., and Woodroof, J. G. (1976). "Commercial Vegetable Processing." Avi, Westport, Connecticut.

Mellor, J. D. (1978). "Fundamentals of Freeze-Drying." Academic Press, New York.

Peterson, C. S. (1979). "Microbiology in Food Fermentations," 2nd ed. Avi, Westport, Connecticut.

Sivitz, M., and Desrosier, N. W. (1979). "Coffee Technology." Avi, Westport, Connecticut.

Stadelman, W. J., and Cotterill, O. J., eds. (1977). "Egg Science and Technology," 2nd ed. Avi, Westport, Connecticut.

Van Arsdel, W. B., Copley, M. J., and Morgan, A. J., Jr., eds. (1973). "Food Dehydration," Vol. 2. Avi, Westport, Connecticut.

Woodroof, J. G., and Luh, B. S. (1975). "Commercial Fruit Processing." Avi, Westport, Connecticut.

Chapter 9

Food Laws and Regulations

Historical Survey

The use of the regulatory powers of government to control the purity, safety, and other attributes of foods entering into commercial trade originated very early in the industrialization of the food industry. For example, early Mosaic and Egyptian laws made provision for how cattle were to be selected and slaughtered for use as food and how the meat was to be handled afterward. Several centuries before the time of Christ, India had regulations prohibiting the adulteration of grain and edible fats and oils. During the Middle Ages in England, strong measures were taken to control adulteration of pepper by such worthless materials as ground nutshells, olive pits, and even iron oxide! In the United States food laws and regulations were enacted by several states as early as the 1780s. (see Chapter 1).

Modern Times

The first serious federal efforts to regulate malpractices in the American food industry did not take place until near the end of the nineteenth century. It took the crusading efforts of Dr. Harvey W. Wiley, chief chemist for the USDA, to make the need for federal action fully appreciated. He brought to the attention of the public and the Congress the deplorable conditions in parts of the food industry of that time, especially the widespread use of such poisonous preservatives as formaldehyde and boric acid. Wiley (1907) published a most interesting and readable account of his attempts to secure adequate federal legislation to protect consumers from fraudulent and dangerous practices. About this same time, Upton Sinclair, a well-known author of the period, published *The Jungle,* which exposed the deplorably unsanitary conditions in the meat packaging industry. Books such as those by Wiley and Sinclair, backed up by a public clamor and a number of concerned legislators, led Congress to enact some much-needed laws in the early part of the present century. The Pure Food and Drug Act and the Meat Inspection Act, signed by President Theodore Roosevelt in 1906, had an immediate and salutary effect on the U.S. food industry, especially in outlawing the use of dangerous preservatives and the trade in meat from diseased livestock, and in bringing about greatly improved sanitation in the processing and preservation of food.

Major revisions in federal legislation pertaining to the safety and purity of most foods were made in the 1930s. The main purpose of the new legislation was to modernize the Food and Drug Act of 1906. Tremendous technical changes and rapid growth had taken place in the food processing industry since the earlier legislation. It was felt that the new conditions called for a more up-to-date and sophisticated approach in drafting laws that would provide adequate protection for the consumer. After several years of public hearings and much debate, the Congress passed and President Franklin D. Roosevelt signed the present Food, Drug and Cosmetic Act of 1938. This legislation, with the several amendments that have been incorporated into the law since, stands today as one of the world's most forward-looking and effective laws relating to food safety and purity and the nutritional value of processed foods. Another law enacted by the Congress, the Fair Packaging and Labeling Act of 1966, filled some of the remaining gaps while avoiding comprehensive legislation in this field. This law requires that retail packages (and their labels) be nondeceptive and contain information of value to the consumer: the name of the food, a list of ingredients, net weight (volume), and name and address of the manufacturer or distributor. Label design and format were also brought under control by this act.

The development of legislation pertaining to the meat and poultry industries followed a somewhat different course than for other foods, primarily because of the special problems that are involved. Livestock and poultry may harbor microbes that cause disease in humans (see Chapter 4). Because of this, the veterinarian in particular is concerned about how to prevent the transmission of these diseases to humans, especially during slaughter and subsequent processing and preservation of meat into food products. The efforts of these professionals and others resulted in the first Meat Inspection Act of 1906 already discussed. In 1957 the Poultry Products Inspection Act was passed. It provides the same type of regulations for the slaughter of poultry and the processing of poultry meat. Legislation governing the egg products industry was passed in 1970. Specific federal legislation pertaining to fish and other foods of aquatic origin has not yet been enacted. However, the Food, Drug and Cosmetic Act and its amendments do provide considerable coverage for processed foods of aquatic origin.

Current U.S. Food Laws and Regulations

Federal statutes dealing with food are administered by a number of different agencies: the Departments of Health and Human Services (USHHS), Commerce (USDC), and the Treasury (USDT); the Environmental Protection Agency (EPA); and the Federal Trade Commission (FTC). This fragmentation of jurisdiction may seem anomalous to the layman, but there are historical, political, and technical reasons for it. Obviously, the arrangement leads to some overlapping of authority and thus an opportunity for serious jurisdictional disputes. However, thus far these have been minimized by the establishment of memoranda of agreements between agencies that spell out the specific authorities and responsibilities of each. In cases of unresolved differences in interpretation of the law, procedural means for resolving them have been established and seem to be effective. For a summary, see Code of Federal Regulations (1966), Hui (1979), and Schultz (1981).

Food and Drug Administration (FDA)

The FDA, a unit of the Department of Health and Human Services, has very broad responsibilities for regulating the safety, purity, and certain other attributes of food and its packaging for that portion of the nation's food supply moving in interstate commerce. The agency's authority stems from the Food, Drug and Cosmetic Act, the Fair Packaging and Labeling Act, and several other statutes, and the amendments that have been made to these laws.

The FDA has developed *standards of identity, standards of quality,* and *fill of container* specifications. These three categories provide the bases on which the FDA judges compliance with its regulations. Standards of identity defines what certain specific foods may contain. For example, strawberry jam is defined as a product that must contain 45 parts by weight of strawberries and 55 parts sucrose, and/or other natural sweeteners such as corn syrup. Another example concerns food bearing the label declaration "enriched." Such foods must contain those vitamins, minerals, or other nutrients in such amounts as FDA stipulates. The term cannot be used when smaller amounts are employed (see Chapter 6).

The official standards of quality specify a minimum quality level for certain foods. For example, for canned pitted cherries the standard specifies the maximum number of defects (pits) that are permitted per unit quantity of fruit. Another example concerns canned cream-style corn. The standard for this product specifies the degree of "spreadability" of the product when tested under certain conditions. (These FDA standards should not be confused with the USDA's voluntary grades for food products, to be discussed later on in this chapter.) The specifications for fill of container state how full the package of food must be to avoid violation for "slack fill."

The FDA has not developed standards of identity or standards of quality for all of the thousands of food items offered for sale at retail. For the present, only about 200 of the more common foods have established federal standards of identity and/or standards of quality.

The FDA also has authority for establishing the types and amounts of chemical additives and other nonnutritive substances that may be found in food. As mentioned in Chapter 4, the FDA agrees with the principle that certain substances may be added to foods provided that 1) they favorably affect some quality attribute of a food and 2) their use does not cause a health hazard. The FDA must review and pass on all applications for the use of such substances in foods. There is available a list of additives that have already been approved for use (see p. 192 for chemical preservatives). It is concerning the need for chemical additives in foods that the FDA, state agencies, and various sections of the public have divergent views. In the authors' opinion there is no excuse for excessive intentional (avoidable) use of chemicals in food products. However, in many cases the intrinsic quality, as well as the keeping quality of a food, may be improved by appropriate chemical additives. The path of wisdom is not always easy to find. In Chapter 11, we will discuss this issue in greater detail.

The FDA also has broad authority for dealing with chemical contaminants in foods, such as pesticide residues and chemicals absorbed from packaging materials. If such a chemical is found to have contaminated the food inadvertently, the manufacturer or other interested party must make application to FDA for a tolerance. Approval may be obtained if it is shown that the contamination is unintentional and unavoidable and that at the permitted tolerance level it does not present a health hazard to humans.

Other types of contaminants, such as foreign and extraneous matter, also come under the FDA's jurisdiction. Also, this agency can condemn products for spoilage or excessive quality deterioration. In addition, products may be seized and destroyed if they are shown to contain microorganisms or their metabolic products which pose health hazards to people. For example, the FDA has strict regulations about the presence of extraneous material in spices, excessive amounts of mold and insect parts in tomato products, and the presence of salmonella organisms in milk, eggs, and other foods. As a matter of fact, a substantial part of the FDA's regulatory activities is devoted to examining foods for such contaminants.

U.S. Department of Agriculture (USDA)

Food Safety and Quality Service (FSQS). The FSQS has full responsibility for administering the Meat Inspection Act and the Poultry Products Inspection Act. Provisions of these laws include ante- as well as postmortem inspection of livestock and poultry for disease and other unwholesome conditions. Also, control is exercised over plant sanitation and other aspects of the safety, purity, and composition of the food products manufactured from "red" and poultry meats, including packaging. The FSQS coordinates its activities with those of the FDA concerning packaging, the deliberate use of food additives, the presence of incidental contaminants in meat and poultry products, and other aspects of product quality and sanitation. The FSQS has almost complete jurisdiction over all matters relating to standards of identity, purity, and safety of meat and poultry products. It has a very special interest in preventing the spread of diseases and intoxications from food animals and their products to humans. For this reason, the agency uses graduate veterinarians in all supervisory positions involving the inspection of live and slaughtered speciments for evidence of disease and for maintaining sanitation.

Agricultural Marketing Service (AMS). The AMS is responsible for administering the Egg Products Inspection Act. The agency also has authority for

Fig. 60. USDA inspection or wholesomeness of broilers. (Courtesy USDA)

voluntary food standards and grades authorized by the Agricultural Marketing Act of 1946. Since the passage of this law, voluntary standards and grades for a wide variety of foods have been established and are in common use at several points in the marketing channel: 1) at the point of first sale (i.e., from the producer), 2) at the wholesale level, and 3) at the retail level. USDA voluntary standards and grades at the retail level are of primary concern here. Consumer grades and standards of the USDA are widely used for "red" meats, chicken, turkey, processed fresh vegetables, and other foods. Ordinarily, only the highest grades (AA and A; Prime, and Choice) of the products appear on the retail markets. Products not meeting specifications for the top grades are usually used for further processing (e.g., for the manufacture of meat pies, sausage, luncheon meats, etc.) The USDA grades and standards emphasize these quality attributes of interest and importance to the consumer. However, safety and purity are not considered in these grades, because to qualify for USDA grading in the first place, the products must have been produced under FSQS-controlled or -approved inspection procedures. The quality attributes AMS uses include appearance and color, texture, tenderness, presence of defects, and uniformity.

The AMS also offers voluntary quality control services for food processors. This is usually called continuous or in-plant inspection and is frequently used where foods are being processed and packed by one food firm for use by another (often called custom-packed). This type of service is in widespread use in the fruit and vegetable freezing and canning industries, and to some extent in others. In these quality control programs plant operations and product grading are under the supervision of an AMS inspector. The packaged foods may or may not show the USDA emblems for continuous inspection and/or grade (Fig. 61), depending on the wishes of the buyer. It is important

Fig. 61. Emblems used in USDA inspection for wholesomeness and quality grading of poultry. (Courtesy USDA)

to emphasize that the FDA, not the AMS, has final jursidiction over the safety and purity of products as sold to consumers.

U.S. Department of Commerce (USDC)

National Marine Fisheries Service (NMFS). The NMFS provides the marine fisheries industry with a number of voluntary services authorized under the Fish and Wildlife Act and the Agricultural Marketing Act. Four of these will be discussed here: 1) in-plant inspection, 2) inspection and grading of products moving in market channels, 3) inspection of plants for sanitation, and 4) technical consultation relating to the foregoing. These services are paid for by the parties requesting them. The in-plant inspection program is comprehensive. A federal inspector supervises all processing operations from raw materials to the packaging of the final product, and also regularly monitors plant sanitation procedures and advises management of any corrective measures that may be required to satisfy NMFS standards.

Grading of the finished product based on NMFS voluntary standards is optional. NMFS grades (A, B, C) are based on a sensory scoring system. Only Grade A products may use the official shield on the package. The plant sanitation service involves a determination of the sanitary conditions in the plant at the time of inspection and not during the actual processing of products. Approved plants are issued certificates indicating compliance with NMFS sanitation standards.

The service providing for the inspection and grading of products already in the market channel calls for elaboration. On request an NMFS grader will sample and grade products at almost any point in the market channel, but most commonly it is done in a cold-storage warehouse. A certificate is issued indicating the grade of the particular lot examined by the NMFS official.

Consultation services are provided to any organization that uses any of the aforementioned services or that wishes to do so. Consultations consist largely of explaining the nature of NMFS services offered, the conditions that must be met in order to qualify, and the costs involved.

Many of NMFS services are used by buyers and brokers of marine fish and shellfish, who find them very valuable in determining the quality of products being offered; this in turn serves as a basis for establishing a fair price to be paid for the product by the buyer or broker. The use of the U.S. grades on the labels of packaged fish and shellfish found in the supermarket is not common.

It is important to note here, as we have for the other voluntary federal programs, that the use of NMFS services does not mean that a firm's products are exempt from FDA regulations. However, it does mean that the FDA is much less likely to find products produced under NMFS inspection programs to be in violation of FDA regulations. Accordingly, these programs are helpful to processors in avoiding difficulty with the FDA.

National Bureau of Standards (NBS). The NBS has sole federal responsibility for the promulgation of official standards of weights and measures for all commodities sold by weight or volume, including foods.

Office of Standards Development Service. This agency is responsible for developing voluntary standards for limiting the classes, sizes, dimensions, and so forth, for a variety of products. Those that affect the food industry are the standards for sizes of packages. To date, only a few have been established: packages for salt and for instant potatoes, containers for green olives, and bags for ice.

U.S. Environmental Protection Agency (EPA)

The EPA has the authority to 1) establish regulations for the deliberate use of pesticides and other chemicals in the environment, including agriculture; 2) establish tolerances for pesticides in food; 3) establish criteria for drinking water (including water used as an ingredient in food); 4) monitor compliance and the effectiveness of surveillance and enforcement; and 5) provide technical assistance to the states.

U.S. Department of Treasury (USDT)

Bureau of Alcohol, Tobacco, and Firearms (BATF). The BATF has broad responsibilities for standards of identity and the purity and safety and other attributes of alcoholic beverages, distilled vinegar, and fruit essences made by a distillation process. In addition, it administers federal regulations relating to the packages and labels used for these products, as well as related advertising and promotion practices. Besides these, the BATF serves as a collecting agency for taxes imposed on alcoholic beverages. This authority is based on several federal statutes: the Internal Revenue Code, the Alcohol Administration Act, The Food, Drug and Cosmetic Act, and Fair Labeling and Packaging Act. Because these are obviously areas of overlap with the authorities of other federal agencies, memoranda of understanding between BATF and other agencies provide means for resolving jurisdictional disputes that arise.

U.S. Federal Trade Commission (FTC)

The FTC enforces the provisions of the Federal Trade Commission Act, which deals with unfair methods of competition and unfair and deceptive trade practices. Most of the emphasis in judging compliance with the law deals with unfair and deceptive advertising and promotion of foods, but not of alcoholic beverages. The FTC coordinates with the FDA in enforcement activities because there is some overlapping of authority, especially with regard to labeling of processed foods.

State and Local Laws and Regulations

Many states, counties, and municipal jurisdictions have laws and regulations dealing with food processing, manufacture, and distribution, including food service in public places. Some states have laws that closely parallel those in force at the federal level. The intention of these laws and regulations is to provide consumer protection for foods produced and distributed solely within the states, where federal regulations usually do not apply. When processors or manufacturers engage in interstate commerce they must comply with existing federal laws and regulations. Most food processing firms of any size do ship their products to other states, and therefore come under federal jurisdiction.

Trade in fresh milk, cream, cottage cheese, sour cream, buttermilk, and yogurt tends to be localized within the states. Thus states have special laws and regulations pertaining to these dairy products, usually patterned after model legislation sponsored by the U.S. Public Health Service. These laws are commonly administered by the state, county, or city health department. The state department of agriculture may also be involved. Sometimes the county agricultural commissioners have some jurisdiction over these products. Enforcement of state, county, and municipal laws and regulations varies widely among the states. However, in some it is just as strict as at the federal level. The State of California is a good example.

International Aspects of Food Laws and Regulations

As early as 1962, a joint food and Agricultural Organization and World Health Organization (FAO/WHO) Conference was held in Geneva, Switzerland, to study international cooperation in food laws and regulations. More than 44 countries sent representatives. As a result, the Codex Alimentarious

Commission was formed in 1963 to deal with the development of international food standards. The commission has held many meetings since that time. Working committees have been set up to develop standards for a number of commodities. Although the negotiations have been complex, tedious, and slow, some progress has been made. One reason for the delay is the extreme variation in food processing practices in the food industries of different nations of the world. Another reason is the desire of local food firms to maintain traditional practices and in some cases to limit food imports or to levy excessive import duties on food.

The USDA, in collaboration with key persons from the U.S. food industry, represents the United States on the Codex Alimentarius Commission. United States representatives have taken an active role in the deliberations to date with the clear intent of moving toward the development of adequate and workable food standards that will protect the interests of consumers and at the same time facilitate trade in foods around the world.

REFERENCES

Code of Federal Regulations (1981). "Title 21 Food and Drugs." U.S. Govt. Print. Off., Washington, D.C.

Hui, Y. H. (1979). "U.S. Food Laws, Regulations and Standards" Wiley, New York.

Schultz, H. W. (1981). "Food Law Handbook." Avi, Westport, Connecticut.

Wiley, H. W. (1907). "Foods and Their Adulteration." McGraw-Hill (Blackiston), New York.

Chapter 10

Careers in Food Science and Technology

Overview

In the universe of science and technology, food science and technology maintains special relationships with several basic disciplines as well as with a number of applied specializations. We depend on the basic principles of the physical, biological, and behavioral sciences in making applications to food processing and food preservation. We also share a concern in providing food for mankind with such specialties as agriculture, fisheries, nutrition, veterinary medicine, business administration, and marketing (see Fig. 62).

Food science and technology itself is composed of several components (see Fig. 63). Discipline-oriented segments include food chemistry, food biochemistry, food microbiology, and sensory analysis; technological segments include engineering, packaging, processing, quality assurance, and sanitation.

Excellent career opportunities in the food industry, related industries, education, and government exist for persons trained in a variety of scientific disciplines and specializations. In this chapter, however, special emphasis is placed on opportunities for those trained more specifically in food science and food technology. Where other specialists work closely with food scientists, they too will be mentioned.

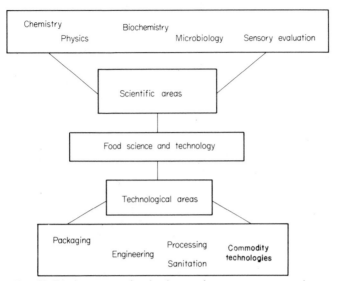

Fig. 62. Food science and technology and its component specialties.

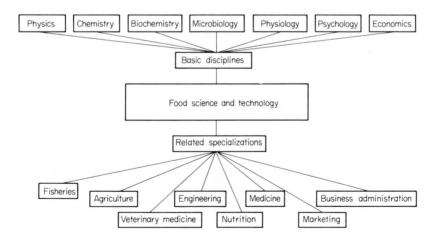

Fig. 63. Food science and technology in the universe of science and technology.

Career Opportunities and Educational Prerequisites

Industry

There is a wide variety of jobs available for young men and women trained in the physical, biological, and behavioral sciences as applied to food processing and food preservation. In the United States the majority of these positions are in the food industry and the related supplier industries (supplying ingredients, packages, services, etc.). The remainder are in colleges and universities, government agencies, and a variety of other organizations concerned with the processing and preservation of food.

Positions in industry are involved in one or more of the following activities: 1) quality assurance, 2) technical service, 3) product development, and 4) process development. The job descriptions and the educational backgrounds required to qualify for these positions are described in what follows.

Quality Assurance. The greatest proportion of the technically trained people working the food industry work in quality assurance (also called quality control). Quality assurance involves an organized control activity, the primary purpose of which is to guarantee that certain levels of quality and other attributes of the processed food have been met. In other words, the entire spectrum of activity from raw material selection and handling to processing, preservation, packaging and finally distribution must be monitored and controlled in order to assure that the food possesses the required characteristics when it reaches the consumer.

In the United States, professionals employed in quality assurance are required to have a B.S. degree, usually in food science. However, it is not uncommon for firms to hire graduate chemists or graduate microbiologists for this type of work. Such employees are usually given special on-the-job training in food quality assurance before undertaking any real responsibility. In any case a new quality assurance employee usually begins by carrying out rather routine assignments, such as collecting samples from the production line and making laboratory examinations of them, or assisting in the analyses, or preparing daily quality assurance reports. As the employee gains competence and experience, she or he usually moves up into a supervisory position and eventually might become director of quality assurance reports. An especially important prerequisite for the job, from a technical point of view, is that the person have a good working knowledge of food chemistry (especially analytical chemistry), food microbiology, and food-plant sanitation. Also, it is essential that she or he understand statistical theory and, even better, understand and be able to use statistical quality control methods. It may be necessary for the person to go back to school or to attend special short courses in order to make up deficiencies.

Something should be mentioned about personal characteristics. Invaluable to success in the field of quality assurance are conscientiousness, systematic and thorough work habits, and an appreciation of the need for precision and accuracy in analytical work. Then too the person needs to be able to communicate effectively with others in the firm, both verbally and in writing. This is so because the success of a quality assurance program depends to a great degree on transmitting technical information to those who need and must use it to achieve control of quality and other attributes of the processed foods produced.

Product Development and Improvement. A substantial number of food professionals work on the development of new or improved products (Fig. 64). Two things are required of the employee in such an activity: 1) creating or acquiring viable concepts for new or improved products, and 2) developing formulations and technologies for the commercial production of these products.

An example of a new-product concept would be a mix for making ice cream at home. Potentially such a product would be popular with consumers, if it is easy to prepare and convenient to use, and if an ice cream of good

Fig. 64. Product-development laboratory. (Courtesy Foremost–McKesson, Inc.)

quality can be produced easily in the kitchen. Developing such a product requires work in the test kitchen, possibly also some research in the laboratory, and certainly some production trials in the pilot plant. Whereas a food scientist would generally head up such a project, at least through the initial stages, she or he usually needs help from a home economist, a technical service-person, and possibly also an engineer. In addition, if special problems in providing adequate shelf life or in solving health hazard problems are encountered, she or he might need the assistance of a chemist and/or a microbiologist.

In the United States, product development departments usually employ one or more persons trained in food science and technology. A minimum of a B.S. degree is essential, and not uncommonly those with M.S. and Ph.D. degrees in food science are preferred. This is particularly true if the firm is large and sophisticated and has a variety of product interests. A product development person needs to have a good working knowledge of food chemistry, biochemistry, and food processing and preservation technologies.

Certain personal characteristics will help make a product development person successful. Above all he or she should be imaginative and resourceful, should enjoy making innovations, and should be readily motivated to create new and improved products. One certainly must be able to sell one's self and one's idea to others, especially to the marketing and top management people in the firm.

Process Development and Improvement. A limited number of professional people working in food firms and in supplier firms are employed in process development and process improvement. These people develop new methods or improve on existing ones for the processing of food. For example, the development of a system and equipment for commercially freeze-drying foods required some entirely new methods of food handling and processing, mainly because of the high vacuum requirement in their production. Also, special problems had to be solved to prevent moisture pickup after freeze-drying and before packaging.

Process development and process improvement projects are usually placed in the hands of graduate engineers. Both chemical and mechanical engineers are qualified for this type of work. Because of their limited knowledge of and experience with food, it is not unusual for them to team up with the food scientist, who is more knowledgeable about the properties and attributes of food. It is considered extremely important for the process development person to have a firm grasp of engineering concepts and of unit operations.

Successful process development and process improvement also calls for certain personal characteristics, in addition to a good academic background: imagination, inventiveness, and resourcefulness, as well as an ability to sell one's self and one's ideas to top management in the firm.

Technical Service. Suppliers to the food industry (firms furnishing ingredients, equipment, sanitation aids, packaging materials, and certain services) employ technically trained personnel to work with customers. Although they may do some selling (and certainly their key role is in support of sales), their main efforts are usually directed at assisting customers in problem solving. For example, a technical service person working for a sugar company would work with a soft drink manufacturer in the selection of the type and amount of sweetener for a certain beverage. Another example would be for a technical service person from a flavor house to advise a meat packing firm in the selection of flavoring materials for a luncheon meat.

The appropriate academic training for a technical service person is a B.S. degree in food science. Not uncommonly, however, graduate chemists are employed for this type of work and given special on-the-job training before being sent out to deal with customers. In either case an effective and successful technical service person needs to have a good working knowledge of food chemistry, biochemistry, and microbiology, as well as some knowledge of the commodity technologies with which he or she works. It is also very valuable to him or her to be familiar with applicable laws and regulations pertaining to processed foods.

Personal characteristics of importance include good communication skills, aggressiveness, effectiveness in "troubleshooting," and, above all effectiveness in working with the variety of people with whom the technical service-person comes into contact.

Management. With the increasing size and technical complexity of the food processing industry, there is a rising demand for management personnel with a technical background. These people share with others trained in marketing, finance, law, and other areas the responsibility for policy making and management in food manufacturing firms. The academic training required for these jobs is at least a B.S. degree in food science plus specialized training in business administration. The latter is often taken after a person has been with a firm a few years and is sometimes financed by that firm. Personal characteristics of importance for management positions include a capacity to analyze and interpret complex and complicated business situations, an ability to work effectively with a variety of people, and a keen sense of the economic potential of one course of action over another.

College and University Training

There are about 60 colleges and universities in the United States and Canada that offer degree programs in food science and food technology. A variety of courses are taught in these institutions: 1) introduction to food science and technology, 2) food chemistry, 3) food biochemistry, 4) food microbiology, 5) food processing and preservation, 6) food engineering, 7) quality assurance, 8) sensory analysis and 9) food laws and regulations, among others. In addition, graduate courses, seminars, and thesis research are offered for credit (see Appendix to this chapter).

The usual educational background required for these academic positions is a Ph.D. degree in food science or in a related discipline (e.g., chemistry, biochemistry, microbiology, chemical or biochemical engineering, nutrition, or psychology). In addition, a number of personal characteristics are important: ability to communicate, a sincere interest in college teaching and in education generally, and the ability to stimulate and motivate students.

Most academic positions involve, besides teaching, basic and applied research and/or extension education work (qualifications for which will be given later).

Basic and Applied Research

A large number of food professionals, especially in universities, government laboratories, and research institutes, are employed to work on basic and applied research problems related to processed food. These programs cover a wide spectrum of subjects including 1) studies on physical and chemical properties of foods, 2) biochemistry of food raw materials and the effects of processing, 3) microbiology of food spoilage, 4) public health microbiology, 5) control of food deterioration and spoilage, and 6) microbial fermentations. To list them all would be an almost impossible task.

The academic training required for a research career is generally a Ph.D. degree in food science or a closely related discipline and possibly a year or two of postdoctoral training. However, it is well known that good research can be and is being done by men and women with only a B.S. degree. The important point to remember is that good researchers should have the scientific background to pursue research in the subject matter of their particular interest. Also they should know how to formulate ideas and to design and carry out meaningful experiments that will answer important questions—thereby advancing the field.

Important personal characteristics for the food researcher include inquisitiveness, imagination, initiative and drive, and tenacity. Also essential is the willingness to complete a research project and to communicate the finds to others, both verbally and in writing.

Regulatory Work

An increasing number of food professionals are working for governmental agencies that administer food laws and regulations. A variety of activities are involved: 1) plant inspections, including collecting products and environmental samples; 2) laboratory examination of collected samples to ascertain adherence to official grades and standards and regulations, watching for evidence of such violations as adulteration or illegal amounts of pesticides, and 3) preparation for and participation in litigation against alleged violators of existing laws and regulations.

In the past, many government agencies employed graduate chemists and microbiologists for food regulatory work; however, increasingly, food scientists are being recruited. In either case, a good working knowledge of food chemistry (especially analytical chemistry), food microbiology, and sanitation is required. Specialized on-the-job training is frequently required of new employees because of the specialized problems that must be faced by agencies engaged in regulatory work.

Certain personal characteristics are important: a systematic and careful approach to identifying violators; precision and accuracy in the taking of samples and in subsequent analytical work performed on them; factual, fearless, and convincing presentation of evidence against violators.

Extension Education

An increasing number of university, government, and industry people are engaged in providing information services and continuing-education programs in food processing and preservation. They might be thought of as "go-betweens," filling the gap between the research scientist and the food scientist in industry and also (in certain cases) between the researcher and the operating personnel in the factory. Their job is to see to it that the information generated through research is put to use. The academic training required for this specialist is generally considered to be a B.S. degree in food science or in a closely related field such as animal science, plant science, or entomology. Increasingly, persons with an M.S. or Ph.D. degree are being recruited to carry out these important educational programs. These people need to know food science sufficiently well to be able to interpret research findings to the working technologist in industry and to the operating people in the plant, so that they will be able to use the information in improving operations. Certain personal characteristics have an important bearing in assuring success in this calling, such as an ability and willingness to serve as a technology-transfer agent and effective verbal and written communication skills.

Currently Available Academic Programs

Food Science

University curricula in this field came into existence in the United States about 50 years ago. The number of schools offering undergraduate and graduate work in food science and technology has increased steadily so that today there are approximately 60 such institutions in the United States (see Appendix to this chapter). Very early in this development the Institute of Food Technologists (IFT) (the professional society of food scientists and technologists in the United States), through its education committee, became active in curriculum development, food-oriented course content, and minimum standards for degree programs. The committee has prepared and published brochures on undergraduate education in food science and technology. Three items are included: 1) IFT-approved minimum standards for an undergraduate curriculum, 2) recommendations regarding required subject matter, and 3) graduate education (see IFT, 1977, 1980). The standards and recommendations have been widely accepted by universities in the United States offering work in this field. The uniqueness of a food science curriculum is in its content of food-oriented courses. The IFT minimum standards require that the following courses be taken by the undergraduate students: 1) food chemistry, 2) food analysis, 3) food microbiology, 4) food engineering, and 5) food processing. All of these courses have science course prerequisites so that they may be taught at an advanced level. For example, the food science student would be expected to complete the following chemistry courses: general and inorganic chemistry, qualitative analysis, quantitative analysis, general organic chemistry, and biochemistry; he or she would also be expected to complete a full year of college mathematics and physics and one course each in general biology and microbiology. Course outlines have been prepared by IFT for all of the required courses mentioned above as well as for a general introductory course. These have been widely adopted by departments offering undergraduate courses in the United States.

Graduate work in food science leading to the M.S. and Ph.D. degrees is offerd by a number of American universities. These programs vary in the type and amount of additional course work required and also whether or not a thesis based on original research is required, especially for the Ph.D. degree. Comprehensive examinations must also usually be taken and passed, and an acceptable thesis filed in the university library (see Albrecht, 1979).

Related Curricula

As has already been mentioned, training in fields other than food science and technology can and does permit a person to have a career in food processing and preservation. Some large food companies and certain governmental agencies and universities perfer to hire persons trained in a related discipline rather than specifically in food science. They believe that they themselves can better provide the orientation to food by special on-the-job training programs during the first few years of employment.

Graduates holding B.S. degrees from the following undergraduate curricula have found employment in the field: chemistry, biochemistry, microbiology, chemical engineering, and mechanical engineering. Holders of M.S. and Ph.D. degrees in the following disciplines and specializations are regularly hired by segments of the food industry, certain government agencies, and some food science departments: chemistry (including agricultural chemistry, analytical chemistry, organic chemistry, and physical chemistry), biochemistry, nutrition, chemical engineering, biochemical engineering, statistics, and psychology.

REFERENCES

Institute of Food Technologists (1977). IFT Undergraduate Curriculum Minimum Standards. *Food Technol. (Chicago)* **31**(10), 60–61.

Albrecht, J. J. (1979). Graduate training for a career in industry. *Food Technol.* (Chicago) **33**, 26–27.

Institute of Food Technologists (1980). "A Career Guidance Brochure." IFT, Chicago, Illinois.

Appendix:
Directory of Universities and Colleges in
United States and Canada Offering 4-Year
Programs in Food Science*

Alabama A and M University
Department of Food Science and
  Technology
Normal, Alabama 35762

University of Alberta
Department of Food Science
Edmonton, Alberta T6G 2N2

University of Arizona
Department of Food Science
Tucson, Arizona 85721

University of Arkansas
Department of Horticulture and Food
  Science
Fayetteville, Arkansas 72701

Auburn University
Food Science Program
Department of Animal and Dairy Science
Auburn, Alabama 36830

Bishop College
Department of Life Science
Dallas, Texas 75241

Brigham Young University
Department of Food Science and Nutrition
Provo, Utah 84602

University of British Columbia
Department of Food Science
Vancouver, British Columbia V6T 1W5

University of California at Berkeley
Department of Nutritional Sciences
Berkeley, California 94720

University of California at Davis
Department of Food Science and
  Technology
Davis, California 95616

*Courtesy of IFT.

California Polytechnic State University
Department of Food Science
San Luis Obispo, California 93401

Chapman College
Department of Food Science and Nutrition
Orange, California 92666

Clemson University
Department of Food Science
Clemson, South Carolina 29631

Colorado State University
Department of Food Science and Nutrition
Fort Collins, Colorado 80521

University of Connecticut
Department of Nutritional Sciences
Storrs, Connecticut 06268

Cornell University
Department of Food Science
Ithaca, New York 14850

Delaware Valley College
Department of Food Industry
Doylestown, Pennsylvania 18901

University of Delaware
Department of Food Science and Human
  Nutrition
Newark, Delaware 19711

Drexel University
Department of Nutrition and Food
Philadelphia, Pennsylvania 19104

Florida A and M University
Foods, Nutrition and Institutional
  Management
Tallahassee, Florida 32307

University of Florida
Department of Food Science and Human
  Nutrition
Gainesville, Florida 32601

University of Georgia
Department of Food Science
Athens, Georgia 30601

University of Guelph
Department of Food Science
Guelph, Ontario N1G 2W1

University of Hawaii
Department of Food Science and
    Technology
1920 Edmondson Rd.
Honolulu, Hawaii 96822

University of Illinois
Department of Food Science
Urbana, Illinois 61801

Iowa State University
Department of Food Technology
Ames, Iowa 50010

Kansas State University
Food Science Program
Manhattan, Kansas 66506

Kansas State University
Department of Engineering Technology
Manhattan, Kansas 66502

University of Kentucky
Food Science Program
Department of Animal Science
Lexington, Kentucky 40506

Université de Laval
Department of Food Science
Quebec, Quebec G1K 7P4

Louisiana State University
Department of Food Science
Baton Rouge, Louisiana 70803

University of Manitoba
Department of Food Science
Winnipeg, Manitoba R3T 2N2

University of Maryland
Food Science Program
Animal Sciences Center
College Park, Maryland 20742

University of Massachusetts
Department of Food Science and Nutrition
Amherst, Massachusetts 01002

Massachusetts Institute of Technology
Department of Nutrition and Food Science
Cambridge, Massachusetts 02139

Michigan State University
Department of Food Science and Human
    Nutrition
East Lansing, Michigan 48823

University of Minnesota
Department of Food Science and Nutrition
St. Paul, Minnesota 55108

Mississippi State University
Food Science Institute
State College, Mississippi 39762

University of Missouri
Department of Food Science and Nutrition
Columbia, Missouri 65201

University of Nebraska
Department of Food Science and
    Technology
Lincoln, Nebraska 68508

North Carolina A and T State University
Department of Home Economics
Greensboro, North Carolina 27411

North Carolina State University
Department of Food Science
Raleigh, North Carolina 27607

Ohio State University
Department of Food Science and Nutrition
Columbus, Ohio 43210

Ohio State University
Department of Horticulture
Columbus, Ohio 43210

Oregon State University
Department of Food Science and
    Technology
Corvallis, Oregon 97331

Pennsylvania State University
Department of Food Science
University Park, Pennsylvania 16802

Purdue University
Food Sciences Institute
West Lafayette, Indiana 47907

University of Rhode Island
Department of Food Science and
  Technology
Nutrition and Dietetics
Kingston, Rhode Island 02881

Rutgers—The State University
Department of Food Science
New Brunswick, New Jersey 08903

University of Tennessee
Department of Food Technology and
  Science
Knoxville, Tennessee 37916

Texas A and M University
Food Science Program
Department of Animal Science
College Station, Texas 77843

Texas Technical University
Agricultural Engineering and Technology
  Dept.
Lubbock, Texas 79409

Tuskegee Institute
Department of Food Science and Human
  Nutrition
Tuskegee, Alabama 36088

Utah State University
Department of Nutrition and Food Science
Logan, Utah 84321

Virginia Polytechnic Institute and State
  University
Department of Food Science and
  Technology
Blacksburg, Virginia 24061

Washington State University
Department of Food Science and
  Technology
Pullman, Washington 99163

University of Washington
Institute for Food Science and Technology
College of Fisheries
Seattle, Washington 98105

University of Wisconsin
Department of Food Science
Madison, Wisconsin 53706

University of Wisconsin
Department of Animal and Food Sciences
River Falls, Wisconsin 54022

Chapter 11

Food and Health Issues—and Answers

In recent years there has been rising concern, especially on the part of the consumer, about the quality of the American food supply. Questions range widely but tend to focus on food/health issues. Attempts to deal with these concerns by various groups have led to a rash of controversies. The answers the consumers are getting to their questions are often contradictory, with knowledgeable scientists generally lined up on one side and self-appointed food/nutrition "experts" on the other. Many consumers, not knowing whom to believe, are distressed and often confused and are left wondering which way to turn for help. Clearly there is a need for a more rational and positive approach to dealing with these consumer concerns and, if possible, to settling the controversies.

The authors believe that knowledgeable and articulate scientists should be playing a more active and positive role in this situation. In particular, we believe that the agricultural scientist, food scientist, nutritionist, and toxicologist could, because of their scientific training and experience, be much more effective in responding to these consumer concerns and in helping to resolve the controversies. In this chapter we will show how the food scientist might serve the public interest in this capacity.

First we will examine in some detail the reasons for consumer concerns about the quality of foods purchased at the market. Next we will show how the controversies concerning food/health issues get started and become perpetuated. Then we will develop a rationale whereby the food scientist might deal with these concerns and thus help to resolve the controversies. Finally, several of the current controversies will be examined and an attempt made to settle them using our rationale.

There are several reasons why food scientists should become involved in dealing with these consumer concerns and the controversies related to them: 1) because they are themselves consumers, they need a sound rationale for judging the quality of their own food; 2) as members of a responsible scientific community, they should be prepared to offer the consumer factual and unbiased information about the quality of commercially processed foods; and 3) if they work for industry, they are in a position to advise their companies concerning advertising and promotion practices and to assure that these are technically accurate, informative, and nondeceptive.

Consumer Concerns and Causes of Controversy

Consumer Concerns

The food scientist should understand the basis for consumers' rising concern about food in relation to their health and that of their families. A number of reasons can readily be identified. A rather obvious one is increased consumer awareness of developments in science, especially in the agricultural sciences, food science, nutrition, and toxicology. The mass media—newspapers, magazines, radio, and TV—carry stories about food and health issues almost daily. Many of these stories focus on health problems associated with the consumption of certain types of foods (e.g., obesity, hypertension, cardiovascular disease, and cancer). This type of story naturally generates a lot of consumer interest and not a little concern, especially for those who consider themselves or members of their family at risk.

Another reason for consumer concern arises from their learning, again through the mass media, that a number of the chemicals used in agriculture and industry are poisonous, some highly so. To their dismay and great concern, they have also learned that, under certain circumstances, residues of these chemicals find their way into the foods they purchase at the market. (They may not be aware that some of these chemicals are essential for life, that the amount of these residues in foods is extremely small, and that permitted residue levels are rigidly limited by government regulations.)

Still another area of consumer anxiety is the increasing awareness of the fact that chemicals are deliberately (and legally) used in the processing and preservation of commercially produced foods. Despite the fact that the use of certain chemicals has been shown to improve the quality of food, the mass media seem to give priority to stories about the hazards that, the "experts" referred to before claim, these chemicals pose to human health.

A reason often overlooked stems from the dramatic shift in the population of the United States during the past century from rural to urban areas. There have been serious consequences from this shift, which bear on consumer concerns about the quality of available food supplies. Formerly, rural families produced much of their own food: they grew grain, fruits, and vegetables, and they kept food-producing animals. They processed and preserved some of the food for future use, either at home or at a nearby commercial food plant, such as the local flour mill, cheese factory, or creamery. With so much of the production and processing under their own control (or that of the "friendly" local processor), these rural people felt quite confident about the quality of the food they served, especially its healthfulness and wholesomeness.

Contrast that with the position of today's urban families. They know little about the modern farming methods that are used for producing crops or food animals. Further, we believe that they know even less about how the raw materials are transformed into the processed food they buy at the market. It is true that if they read the feature stories in newspapers and magazines or listen to or look at specials on radio and TV, these city dwellers could learn a good deal about modern agriculture and current food-processing methods. However, these media events are usually considered entertaining but only vaguely educational. They do not seem to provide consumers with the same assurance of quality in their foods that was enjoyed by their parents or grandparents. Then, too, today's consumers can read the labels on processed and packaged foods and thereby learn what is in them and even the RDAs they supply. Label information is not always reassuring to the uninitiated, however, especially when they see terms like butylated hydroxyanisole (BHA, an antioxidant) or trisodium phosphate (TSP, a pH control agent). Most consumers have probably never heard of either of these chemicals or wonder why in the world they have been put in their food.

Controversies about Food and Health

As already stated, controversies about food and health abound in this country. It will be worthwhile to trace their origins. Many controversies orginate with the dubious claims made by the food/nutrition "experts." For example, there are the claims made for "organic" food, which we will discuss later. Proponents of these claims frequently have little or no scientific training and experience in the subject matter involved. Certainly most have no standing among reputable scientists active in the field.

Dubious claims are sometimes also made by those somewhat better trained in science. Examples are claims of serious health hazards allegedly caused by the legally approved pesticide residues or the chemical additives used in processed food. Some of the critics have degrees in chemistry, microbiology, or certain of the so-called health sciences, such as dentistry, nursing, or chiropractic care. Generally speaking, however, they have had little or no training in the relevant subject matter involved in dealing with food/health issues. Even physicians, unless they received their medical training recently, have had only limited training in nutrition and toxicology and almost certainly none in the agriculture sciences or food science. It is the authors' view that, without such scientific training and without experience in these fields, an individual has very limited qualifications to make or judge claims about food in relation to health. (Later we will deal more specifically with who is qualified.)

To us it seems fair to blame mass media for their part in perpetuating unproved claims. This is so because of their propensity for featuring the sensational, a practice that plays into the hands of the food/nutrition "expert," who is ready and willing to provide a story. Knowledgeable scientists tend to shy away from dealing with these controversies in their relationships with the media for fear of being misquoted. In all fairness, we are pleased to see that recently some of the media are acting more responsibly in dealing with food/nutrition controversies.

Public relations and advertising agencies have not always acted responsibly in dealing with food/health issues. For example, in recent years they have grossly misused such terms as "natural" and "no chemical preservatives" in food advertising. The result is that these terms have become almost meaningless. Stewart and Mattson (1978) strongly criticized these practices and proposed reforms, but their efforts seem to have had no effect on food advertising/promotion practices to date.

Dealing with the Issues

The authors believe that the food scientist should develop a rationale to use as a guide in dealing with consumer concerns and controversies regarding food in relation to health. We have given the matter considerable thought and study and will present our rationale for dealing with these issues. Four aspects of our proposal will be discussed: 1) life-style considerations, 2) emotionalism as a problem, 3) reliance on science for the facts, and 4) reliance on qualified scientists for the interpretation of research findings.

Life-Style—An Issue or Nonissue

By life-style we mean the typical way individuals conduct their daily lives. Life-styles vary tremendously, but some are affected by formal affiliations:

1) belonging to a religious or philosophical sect, 2) being a member of a formal group concerned with food (e.g., "organic" food movement), or 3) being a member of some type of cult or sect that imposes constraints on food choices and nutritional practices.

The authors adhere to the belief that each individual is entitled to his or her chosen life-style, even if it does impact adversely on the person's health. At the same time, we believe that everyone should try to understand the implications of one's life-style on health. In the present discussion we are talking about the implications of food choices and nutritional practices. In some cases it should be possible to make certain accommodations without violating the tenets of the sect or cult and thus avoid the health problems. Of course, this will not always be possible.

Emotionalism and Irrational Behavior

Closely related to the consideration of life-style is the problem of emotionalism. It is commonly said that humans are the only rational members of the animal kingdom. However, within the context of the present discussion a more precise and meaningful definition would be that humans are the only members of the animal kingdom possessing the capability for rational behavior. The difference is obviously considerable, for although we have this capability, very often our behavior is irrational. Thus we can have all the facts about how certain food choices will affect our health, and still choose to ignore them and take an irrational course, driven by powerful emotional forces we seem not to be able to control. So it is with the mystics who have been informed about the adverse consequences of following the Zen macrobiotic diet (a predominantly vegetarian diet that places great emphasis on whole-grain cereals and the avoidance of fluids). A number of people who have insisted on rigidly following the diet have died of malnutrition (Nutrition Foundation, 1974).

Food scientists must understand the importance of life-style and emotional behavior considerations in food/health controversies. Although they can present the facts and hope for a rational reaction from those with whom they are communicating, there will be times when it will not happen. There is no way one can settle a controversy with people whose behavior is based primarily on emotionalism.

Reliance on Science for the Facts

Only scientific inquiry will reveal the facts needed to judge the pros and cons concerning food choices and nutritional practices in relation to health. Thus it will be instructive to examine the state of knowledge in the several scientific fields that are concerned with food/health issues. What do we know and what remains to be discovered?

Agricultural Sciences. The agricultural sciences embrace, among others, plant science, animal science, aquaculture, entomology, and plant pathology. These specialties have grown rapidly over the past century, both in research output and in scientific stature. In fact, developments in these and other agricultural specialties have provided the basis for the tremendous increase in agricultural productivity in this country during this period. However, certain problems have arisen. In an attempt to make effective use of man-made chemicals (e.g., fertilizers, pesticides, plant-growth regulators, drugs, and pharmaceuticals), agricultural scientists may have neglected to consider fully the adverse effects these chemicals might have on worker safety, food quality, and ecology. Fortunately, in recent decades these and other scientists have increasingly taken up these aspects and have studied methods for eliminating harmful side effects or at the very least minimizing them. Still more research is needed along these lines.

Food Science. As we have seen in previous chapters, the field of food science has grown spectacularly over the past few decades, both in size and stature. In fact, the modern food processing industry is based to a very considerable extent on developments made in food science over the past 150 years. Notwithstanding, in earlier years certain problem areas seem to have been neglected. Insufficient attention was probably paid to establishing the safety of additives deliberately used in food and to the adverse effects of processing and preservation methods on nutritional values. Fortunately, this situation has changed markedly in recent years, with considerable attention now being paid to these problems. Then, too, food scientists have learned to team up with nutritionists and toxicologists in order to solve some of the more complex problems of food safety and nutrition. We think that the future looks bright for discovering all the facts needed to settle controversies.

Nutrition. As we have seen in Chapter 6, there has been tremendous progress in the science of human nutrition over the past 50–75 years. Today we know most, if not all, of the essential nutrients and have a considerable store of knowledge about daily requirements. However, some of the more complex problems that have a nutritional aspect are just now being seriously studied, especially obesity, hypertension, cardiovascular disease, and cancer. Fortunately, great research emphasis is currently being placed on the role of nutrition in preventing and treating these disorders. This research should provide the facts we need to settle some of the controversies mentioned here.

Toxicology. The field known as toxicology had its origin largely in pharmacology, which deals with the beneficial as well as adverse side effects of drugs on humans and other animals. Toxicology has now achieved a status of its own; it deals mostly with the effects of exposure to chemicals and environmental agents on the health of humans and other animals. Although considerable progress has been made in toxicological research, the field is beset by serious methodological problems. Traditionally, short-term (acute toxicity) and long-term (chronic toxicity) animal studies have provided the information that forms the basis for establishing human toxicity. Now, however, a number of toxicologists are seeking more specific and refined methods. This is especially true for agents suspected of causing cancer or behavior disorders (e.g., hyperactivity in children; Conners, 1980). Another methodological problem relates to more firmly determining true toxicity to humans. At present, human toxicity is estimated by the extrapolation of the results of experimental animal studies to humans. (To some extent epidemiological findings are also used, but the results of such studies usually do not permit the establishment of a precise cause-and-effect relationship.)

Fortunately, extensive research is being done on methodological problems in the United States and a number of other countries. If all goes well, within a few years we will have adequate information on the toxicity to humans of chemicals and other agents with which they come in contact. It is essential that this dilemma be solved if we are to have the facts we need to settle current controversies about food safety.

Reliance on Qualified Scientists

The lay person is not sufficiently well trained or experienced to interpret research findings relevant to the food and health controversies. Such interpretation requires well-trained and experienced scientists in one or more of the relevant disciplines: nutrition, toxicology, food science, and the agricultural sciences. To evaluate the claims proponents make about food and health, the lay person also needs the help of persons with demonstrated objectivity and fairness in dealing with these topics. This is sometimes a difficult combination to find, for the latter two personal characteristics are not found in every person, even scientists. Still, it is the authors' belief that there are scientists with both good professional and fine personal characteristics who are available and willing to serve in the public interest.

What are the qualifications for a person to judge the claims made by proponents of certain food choices and dietary practices? Consider the claim that certain food additives, such as antioxidants, present a hazard to the health of consumers. We believe the following qualifications are essential:

1) postgraduate study in toxicology or pharmacology (to the Ph.D. level or the equivalent); 2) several years of research experience in food toxicology, with papers published in peer-reviewed journals; 3) recognition of one's stature in toxicology by scientific peers; 4) no economic interest in the sale or use of the additives; and 5) a record of objectivity and fairness in dealing with controversial issues. In dealing with antioxidant use in food, the toxicologist needs to team up with the food scientist, who is knowledgeable about the benefits derived from the use of these compounds in food. Similar criteria could be developed for scientists to evaluate claims dealing with other subjects about which there is controversy.

These criteria may seem so strict and demanding that few, if any, scientists could qualify. However, as we have just stated, there are some who do qualify and they are ready and willing to serve. A more serious problem for the food scientist is to find the time and make the effort to locate qualified scientists and then to get their views on controversial issues. Fortunately a solution to this problem has been proposed and is in use by the scientific community serving the food field. Thus several scientific bodies have established peer groups or have named qualified scientists to study and report on the controversies about food in relation to health. These people are selected on an interdisciplinary basis so as to get a representation of the various scientific specialties involved in specific controversies. The following are active at the present writing: the Expert Panel on Food Safety and Nutrition of the Institute of Food Technologists (IFT), the Council on Agricultural Science and Technology (CAST) (sponsored by a number of societies representing the agricultural sciences), the Council on Food and Nutrition of the American Medical Association (AMA), and the Committee on Chemistry and Public Affairs of the American Chemical Society (ACS). These organizations and others have prepared a number of position papers on controversies dealing with food in relation to health. These have been well received by the scientific community and by many people in the mass media. It thus appears to be a valid approach to settling the controversies, at least to the satisfaction of most consumers.

A Look at Some Current Controversies

The student will find it interesting and instructive to examine some of the current controversies and then to note how the authors attempt to settle or deal with them. In doing so we will take into account those considerations mentioned earlier: life-style, emotionalism, dependence on science for the facts, and dependence on scientists for the interpretations. Three controversies will be examined from a scientific standpoint: "organic" foods, "natural" foods, and pesticide residues in food.

Organic Foods

A lively controversy has been raging for years about the virtues of organic foods. (The term *organically grown* would be, technically speaking, a more appropriate description. However, "organic" is so widely used and accepted that we have chosen to use it here.) Organic crops are grown without the use of man-made chemicals, such as commercial fertilizer, pesticides, and plant hormones; organic foods of animal origin are produced without the use of drugs, antibiotics, hormones, pesticides, or other man-made chemicals.

Proponents of organic foods and organic farming make a variety of claims for the farming system and the foods so produced: ecological benefits, energy conservation, soil and water conservation, and superior-quality foods. We will concern ourselves here only with the claim that organic foods possess superior nutritional and better sensory qualities and are free of pesticide residues.

The IFT Expert Panel and the Council for Agricultural Science and Technology (CAST) have both published position papers on organic foods (IFT, 1974b; CAST, 1980). In addition, the Nutrition Foundation has published two articles dealing with the claims (Nutrition Foundation, 1974). In all three cases, the papers dispute the claim for superior nutritional values for these foods, citing the results of numerous research studies as evidence.

CAST disputes the claim for better sensory properties. However, in the present author's judgment, the results of the research cited are flawed due to a faulty experimental design. Neither the IFT panel nor the authors of the Nutrition Foundation papers attempted to deal with this claim. Obviously, further research is needed to settle the validity of the claim for better sensory properties.

The claim that organic foods are free of pesticide residues appears to be invalid. Studies show them, at least on occasion, to contain residues, no doubt due to accidental contamination (Barrett, 1980). We will discuss the problem of whether such contamination poses a health problem to the consumer in the section on Pesticide Residues in Food.

For further information on the organic food controversy, the student is referred to the references and reading list at the end of the chapter.

Natural Foods

The controversy about natural foods continues unabated after many years. Sometimes these foods have been considered to be the same as organic foods. However, the terms *organic* and *natural* as applied to foods have very different meanings. Whereas organic refers to the farming methods used in their production, natural refers to the processing and preservation methods employed in their manufacture. It is generally agreed that natural foods are produced using minimal processing and are not processed using certain chemical additives, that is, artificial colors or flavors or man-made chemical preservatives.

Proponents of natural foods claim greater safety and superior nutritional values for them. They state that minimal processing substantially reduces the amount of nutrients extracted from or destroyed in the food during manufacture and that the avoidance of artificial colors and flavors and chemical preservatives provides safer foods. No single position paper was located by the authors that deals broadly with this controversy. However, numerous papers and book chapters dealing with one or more of the claims were found (Barrett, 1980; Labuza, 1977; IFT, 1974a, 1975, 1976, 1980; Deutsch, 1976; Conners, 1980).

There is good evidence that processing can and in certain cases does extract or destroy nutrients (see Chapter 6). At the same time we know that processing can be beneficial, both to nutritional values and safety (see Chapters 1, 4, 6, and 7). For example, processing is used to inactivate or destroy nutritional inhibitors and other toxic materials, and heat treatment is a very effective means for destroying pathogenic microbes. Thus minimal processing may not be the best method for every situation. What is required is to find a balance that maximizes the benefits of processing while minimizing adverse effects. That is precisely what the food scientist strives to do in selecting processing methods.

The chemical additives legally permitted in this country, including nearly all artificial colors and flavors and chemical preservatives, have undergone extensive toxicological testing for safety before regulatory approval. In addition, it is legally required that before approval they be shown to serve a useful purpose in the control of quality. Moreover, the way in which they are used and the amounts remaining in the finished products are rigidly controlled by regulatory agencies (see Chapter 9).

Still, as we discussed earlier (p. 259), toxicology is a young and as yet somewhat immature science. So, there are doubts in some scientific quarters about whether the test methodologies being used are truly adequate, especially those for chemicals suspected of causing cancer or psychological disorders. As test methods have been improved over the years, some additives have been removed from the approved list. It is conceivable that some additional additives may face the same fate as research on improved methodology develops. Time will tell.

In this situation of some uncertainty, the FDA and the USDA have taken very conservative positions in regard to what additives may be used. Thus, we can make use of the approved additives' desirable properties in processed foods without any apparent risk. Almost all food toxicologists agree with the position that these agencies have taken in permitting the use of a limited number of additives in food processing.

Pesticide Residues in Food

Ever since the book, *Silent Spring,* by Rachel Carson, appeared in the 1960s, there has been controversy about the use of pesticides in agriculture, forestry, public health programs, and other areas. The objections to their use are many, including ecological damage, user safety, and food contamination. Here we will deal only with the latter.

There is no doubt that chemical pesticides are poisonous, some exceedingly so, and the residues of those used in agriculture sometimes find their way into the food consumers buy in the market. However, there are very strict regulations concerning their use, and the amount of residue, if any, found in foods sold at retail is severely limited.

The toxicity of these chemicals has been studied exhaustively by toxicologists in industry, universities, and government laboratories. "No-effect levels" (intake levels at which no effects are observed in test animals) have been determined for them. Permitted residue levels in food are set far below these no-effect levels, usually at only a few parts or only a fractional part per million of the food involved. All of this seems to provide adequate protection for the consumer of these foods.

However, as we have noted (p. 262), there are questions about what are appropriate toxicological methods, especially for chemicals suspected of causing cancer. This uncertainty led to the passage in 1962 of the Delaney Clause, an amendment to the Food, Drug and Cosmetic Act. This clause reads: "That no additive shall be deemed to be safe if it is found to induce cancer when ingested by man or animal or if it is found, after tests which are appropriate for the evaluation of the safety of food additives, to induce cancer in man or animal." This clause makes it illegal for any processed food to contain *any amount* of a man-made chemical shown to cause cancer when tested on humans or animals, no matter how high the dose level administered. (An interesting sidelight of this problem is that "zero" residue has become difficult to define because current analytical methods for several pesticides detect less than one part per billion of food! Thus zero seems to be a goal ever more difficult to reach!) Moreover, some metals that are essential to human life are essential parts of some pesticides (copper and zinc are examples).

Faced with all of these problems, government agencies have taken an extremely conservative position in deciding what pesticide residues (or additives, as we have already noted), if any, are allowed in food and, if so, at what levels. However, most toxicologists agree with the official national and international positions and with the current regulations that allow only certain pesticides to be used for the production of foods, as well as with what the minimal residue levels of pesticides in foods sold in the market should be.

Summing Up

It is obvious that consumers have a number of concerns about the quality of their food supply, especially its healthfulness. Indeed, controversies have developed as various groups try to deal with those concerns. In this chapter we have reviewed these concerns, their origins, and reasons for their existence. We have also examined the controversies that have developed, including some of the issues and the positions of the pro and con adversaries.

The authors encourage food scientists to become aware of these concerns and the related controversies and to become active in helping consumers make sound and rational decisions about their food choices and dietary practices. To aid them in this effort we have presented a rationale for dealing with concerns and the related controversies. We urge that consideration be given to several of the important factors involved: 1) life-style, 2) emotionalism, 3) reliance on science for the facts, and 4) reliance on scientists for interpreting the facts. If food scientists' succeed in these efforts, rational consumers will be able to make sound decisions about food choices and nutritional practices, and thus ensure good nutrition and good health for themselves.

REFERENCES

Barrett, S., ed. (1980). "The Health Robbers." Sticley, Philadelphia, Pennsylvania.

Council for Agricultural Science and Technology (1980). "Organic and Conventional Farming Compared," Rep. 84. CAST, Ames, Iowa.

Conners, C. K. (1980). "Food Additives and Hyperactive Children." Plenum, New York.

Deutsch, R. M. (1976). "Realities of Nutrition." Bull Publ., Palo Alto, California.

Institute of Food Technologists (1974a). "Effects of Food Processing on Nutritional Values." IFT, Chicago, Illinois.

Institute of Food Technologists (1974b). "Organic Foods. Scientific Status Summary." IFT, Chicago, Illinois.

Institute of Food Technologists (1975). "Sulfites as Food Additives." IFT, Chicago, Illinois.

Institute of Food Technologists (1976). "Diet and Hyperactivity: Any Connection?" IFT, Chicago, Illinois.

Institute of Food Technologists (1980). "Food Colors," Scientific Status Summary. IFT, Chicago, Illinois.

Labuza, T. P. (1977). "Food and Your Well-Being." West Publ., St. Paul, Minnesota.

Nutrition Foundation (1974). "Nutrition Misinformation and Food Faddism," Nutrition Review Special Supplement. Nutr. Found., New York.

Stewart, G. F., and Mattson, H. W. (1978). Food Advertising and promotion: a plea for change. *Food Technol. (Chicago)* **32**(11), 30–33.

SELECTED READINGS

American Chemical Society (1980). "Chemistry and The Food System—A Study by the Committee on Chemistry and Public Affairs." Am. Chem. Soc., Washington, D.C.

Council for Agricultural Science and Technology (1981). "Food Questions and Answers," Spec. Publ. No. 7. CAST, Ames, Iowa.

Council for Agricultural Science and Technology (1981). "Regulations of Potential Carcinogens in the Food Supply. The Delaney Clause," Rep. 89. CAST, Ames, Iowa.

Clydesdale, F., ed. (1979). "Food Science and Nutrition." Prentice-Hall, Englewood Cliffs, New Jersey.

Institute of Food Technologists (1978). "The Risk/Benefit Concept Applied to Food." IFT, Chicago, Illinois.

Labuza, T. P. (1975). "The Nutrition Crisis—A Reader." West Publ., St. Paul, Minnesota.

Author Index

Italicized entries refer to pages containing complete reference citations.

A

Accum, F. C., 20, *30*
Adam, W. B., *175*
Adams, R. N., 70, *84*
Adamson, J. D., 61, *62*
Akesson, C., 125, *128*
Allen, D. E., *62*
American Chemical Society, 21, 25, *30, 265*
Amerine, M.A., *85,* 103, 119, 121, 125, *126,* 227, *229*
Anonymous, 26, *30,* 42, 43, *62*
Arnott, M.L., 15, 16, *30*
ASHRAE, 206, *229*
Ashley, W., 22, *30*
Ayres, J.C., *101,* 181, *199*

B

Barrett, S., 261, 262, *264*
Bartoshuk, L.M., 111, 114, 115, *126*
Beaton, G.H., 153, *174*
Beets, M.G.J., 108, 109, *126*
Bender, A.E., 168, 169, 170, 171, 172, *174, 175*

Bengoa, J.M., 45, *63*
Bennett, M.K., 42, 57, *63*
Bernard, R.A., 9, *31*
Beuchat, L.B., *101*
Beuk, J.F., *175*
Bigelow, W.D., 191, *199*
Bigwood, E.J., 48, *63*
Bimbenet, J.J., 187, *199*
Birch, G.G., *85,* 106, 107, 113, *126*
Biswas, M.R. and A.K., 40, *63*
Blakeslee, A.F., *85,* 114, *126*
Blaylock, J., 55, *64*
Bodenheimer, F.S., 76, *84*
Bökönyi, S., 3, *30*
Bonnichsen, R., 2, *32*
Borgstrom, G., 41, 43, *63*
Bowers, R.H., 126, *127*
Brand, J.G., 69, *84*
Brennan, J.G., *85,* 106, 107, 116, *126*
Brennen, J.C., *102*
Brooks, R.R.R., 42, *63*
Brothwell, D. and P., 2, *30*
Brown, L.R., 38, *63*
Bryan, F.L., 93, *101*
Burk, M.C., 49, *63*

Subject Index

A

Acetic acid (and acetates), 113, 198, 211, 212, 213; *see also* Vinegar

Acetification, 4, 8, 11; *see also* Vinegar

Acetylcholin, 150

Acid (and acidity), 206, 225, 226; *see also* Acetic, Citric, Lactic, Malic, pH, Propionic, and Sour

Acorn, 71

Acrid, 111

Adaptation, 109 – 110, 114, 116, 117

Additives, 66, 233, 234, 256, 258, 259, 260, 262, 263

Adenosine triphosphate (ATP), 139

Adenovirus, 93

Adulteration, 15, 19 – 20, 183, 230, 247; *see also* Sophistication

Aerobic

Aflatoxin, 92

Agency for International Development, 34

Agglomeration (as a food process), 209

Aging (and maturation), 121, 227 – 228

Ahimsa, 75

Air-sacculitis, 206

Alcohol, 23, 62, 120, 226; *see also* Fermentation

ethyl, 10, 16, 117, 210, 211

Ale, 18, 19; *see also* Beer and Malt

Alfalfa, 15

Alkaline, 111

Alkaloids, 113

Allergens, 90, 120

Almonds, 106, 162

Aluminum, 184, 202, 225

American (cheese), 212

American Medical Association (AMA), Council on Food and Nutrition, 260

American Chemical Society (ACS), Committee on Chemistry and Public Affairs, 260

Amino acids (and amines), 46, 47, 119, 134, 140 – 143, 145, 149, 150, 151, 180, 187

β-Aminopropionitrile, 89

Ammonia, 108

Amylase, 139, 210

Anaerobic, 194

Anemia, 145

Animals, as food consumption, 52, 53, 54; *see also* Beef, Pork, Poultry, Sheep, Turkey, etc.

nutrition, 205

Anthocyanins, 180 – 181, 187

Antibiotics, 79, 90

FOOD SCIENCE AND TECHNOLOGY

A SERIES OF MONOGRAPHS

In preparation

Malcolm C. Bourne (ed.), FOOD TEXTURE AND VISCOSITY: CONCEPT AND MEASURE-MENT. 1982.

Héctor A. Iglesias and Jorge Chirife, HANDBOOK OF FOOD ISOTHERMS: WATER SORPTION PARAMETERS FOR FOOD AND FOOD COMPONENTS. 1982.

John A. Troller, SANITATION IN FOOD PROCESSING AND SERVICE. 1983.